ENERGY POLICIES, POLITICS AND PRICES

OIL AND GAS DEVELOPMENT ON FEDERAL AND INDIAN LANDS

OVERSIGHT AND PRODUCTION ISSUES

ENERGY POLICIES, POLITICS AND PRICES

Additional books in this series can be found on Nova's website under the Series tab.

Additional e-books in this series can be found on Nova's website under the e-book tab.

ENERGY POLICIES, POLITICS AND PRICES

OIL AND GAS DEVELOPMENT ON FEDERAL AND INDIAN LANDS

OVERSIGHT AND PRODUCTION ISSUES

AARON SPEARING
EDITOR

New York

Copyright © 2014 by Nova Science Publishers, Inc.

All rights reserved. No part of this book may be reproduced, stored in a retrieval system or transmitted in any form or by any means: electronic, electrostatic, magnetic, tape, mechanical photocopying, recording or otherwise without the written permission of the Publisher.

For permission to use material from this book please contact us:
Telephone 631-231-7269; Fax 631-231-8175
Web Site: http://www.novapublishers.com

NOTICE TO THE READER

The Publisher has taken reasonable care in the preparation of this book, but makes no expressed or implied warranty of any kind and assumes no responsibility for any errors or omissions. No liability is assumed for incidental or consequential damages in connection with or arising out of information contained in this book. The Publisher shall not be liable for any special, consequential, or exemplary damages resulting, in whole or in part, from the readers' use of, or reliance upon, this material. Any parts of this book based on government reports are so indicated and copyright is claimed for those parts to the extent applicable to compilations of such works.

Independent verification should be sought for any data, advice or recommendations contained in this book. In addition, no responsibility is assumed by the publisher for any injury and/or damage to persons or property arising from any methods, products, instructions, ideas or otherwise contained in this publication.

This publication is designed to provide accurate and authoritative information with regard to the subject matter covered herein. It is sold with the clear understanding that the Publisher is not engaged in rendering legal or any other professional services. If legal or any other expert assistance is required, the services of a competent person should be sought. FROM A DECLARATION OF PARTICIPANTS JOINTLY ADOPTED BY A COMMITTEE OF THE AMERICAN BAR ASSOCIATION AND A COMMITTEE OF PUBLISHERS.

Additional color graphics may be available in the e-book version of this book.

Library of Congress Cataloging-in-Publication Data

ISBN: 978-1-63321-779-9

Published by Nova Science Publishers, Inc. † New York

CONTENTS

Preface vii

Chapter 1 Oil and Gas: Updated Guidance, Increased Coordination, and Comprehensive Data Could Improve BLM's Management and Oversight 1
United States Government Accountability Office

Chapter 2 U.S. Crude Oil and Natural Gas Production in Federal and Non-Federal Areas 41
Marc Humphries

Chapter 3 Oil and Gas Lease Utilization, Onshore and Offshore, Updated Report to the President 55
U.S. Department of the Interior

Index 77

PREFACE

Federal and state regulatory agencies manage and oversee the development of federal and Indian oil and gas resources by issuing rules and guidance, reviewing drilling applications, and inspecting wells to ensure compliance with environmental, safety, and other regulations. In fiscal year 2013, federal and Indian energy production, including oil and gas, generated almost $15 billion in revenue. Recent advances in drilling technologies have greatly expanded the ability to develop oil and gas resources, particularly from shale and tight sandstone rock formations. This book discusses the changes to federal and selected state agencies' rules over the past 5 years and examines the effectiveness of BLM's management and oversight of federal and Indian resources. It also examines U.S. oil and natural gas production data for federal and non-federal areas with an emphasis on the past five years of production.

Chapter 1 - Federal and state regulatory agencies manage and oversee the development of federal and Indian oil and gas resources by issuing rules and guidance, reviewing drilling applications, and inspecting wells to ensure compliance with environmental, safety, and other regulations. In fiscal year 2013, federal and Indian energy production, including oil and gas, generated almost $15 billion in revenue. Recent advances in drilling technologies have greatly expanded the ability to develop oil and gas resources, particularly from shale and tight sandstone rock formations.

GAO was asked to review oversight of federal and Indian oil and gas resources. This report (1) discusses the changes to federal and selected state agencies' rules over the past 5 years and (2) examines the effectiveness of BLM's management and oversight of federal and Indian resources. GAO reviewed federal and selected state agencies' rules and guidance, data from federal agencies, and other documentation. GAO selected a sample of 14 states based, in part, on their involvement with oil and gas development. This information cannot be generalized to all states.

Chapter 2 - In 2013, the price of oil averaged $98 per barrel (West Texas Intermediate spot price), up from $94 per barrel in 2012. Prices remain high in early 2014 (near $100 per barrel) and are projected by the Energy Information Administration (EIA) to average in the mid-$90 per barrel range through 2014. A number of proposals designed to increase domestic energy supply, enhance security, and/or amend the requirements of environmental statutes are before the 113th Congress. A key question in this discussion is how much oil and gas is produced in the United States each year and how much of that comes from federal versus non-federal areas. Oil production has fluctuated on federal lands over the past five fiscal years but has increased dramatically on nonfederal lands. Non-federal crude oil production has been rapidly increasing in the past few years partly due to favorable geology and the relative ease of leasing from private parties, rising by 2.1 million barrels per day (mbd) between FY2009-FY2013, causing the federal share of total U.S. crude oil production to fall by nearly 11%.

Natural gas prices, on the other hand, have remained low for the past several years, allowing gas to become much more competitive with coal for power generation. The shale gas boom has resulted in rising supplies of natural gas. Overall, annual U.S. natural gas production rose by about four trillion cubic feet (tcf) or 19% since FY2009, while production on federal lands (onshore and offshore) fell by about 28%. Natural gas production on non-federal lands grew by 33% over the same time period. The big shale gas plays are primarily on non-federal lands and are attracting a significant portion of investment for natural gas development.

The number of producing acres may or may not be a function of how many acres are leased, and the number of acres leased may or may not correlate to the amount of production, but in recent years, some members of Congress have proposed a $4/acre lease fee for non-producing leases. This proposal grew out of the efforts to open more public land and water (offshore) for oil and gas drilling and development when gasoline prices spiked in 2006-2008. Some in Congress noted that there were many leases they believed were not being developed in a timely fashion, while at the same time, others in Congress were pushing for greater access to areas off-limits (such as the Arctic National Wildlife Refuge (ANWR) and areas under leasing moratoria offshore). Higher rents for offshore leases were imposed by the Secretary of the Interior in 2009 to discourage holding unused leases and to move more leases into production, if possible.

Another major issue that Congress may seek to address is streamlining the processing of applications for permits to drill (APDs). Some members contend that this would be one way to help boost energy production on federal lands. After a lease has been obtained, either competitively or non-competitively, an APD must be approved for each oil and gas well. Despite the new timeline for review (under the Energy Policy Act of 2005, P.L. 109-58), it took an average of 307 days for all parties to process (approve or deny) an APD in 2011, up from an average of 218 days in 2006. The difference, however, is that in 2006 it took the Bureau of Land Management (BLM) an average of 127 days to process an APD, while in 2011 it took BLM 71 days. In 2006, the industry took an average of 91 days to complete an APD, but in 2011, industry took 236 days. The BLM stated in its FY2012 and FY2013 budget justifications that overall processing times per APD have increased because of the complexity of the process.

Chapter 3 - On March 11, 2011, President Obama directed the Department of the Interior (Department) to determine the acreage of public lands that had been leased to oil and gas companies but remain undeveloped, noting that companies should be encouraged to produce energy from leases that they are holding. The Report reached several important conclusions: First, the Department has offered substantial acreage for leasing and resource development, but much of this acreage has not been leased by industry. Second, tens of millions of acres that are currently under lease remain idle. Because these areas are not undergoing exploration, development, or production, taxpayers are not getting the full advantage of America's resource potential.

Soon after the release of the Report, the President released a *Blueprint for a Secure Energy Future,* a comprehensive energy strategy to secure America's energy future by producing more conventional energy at home while working to reduce our dependence on oil by leveraging cleaner, alternative fuels and greater efficiency. Included in the *Blueprint* were a number of steps to encourage safe and responsible domestic energy production both onshore and offshore – including steps to encourage diligent development.

This Report updates the Department's 2011 Report, outlines some of the policies that are being implemented, consistent with last year's findings and the *Blueprint*, and describes some of the geographic areas that best reflect the promise of this Nation's onshore and offshore energy resources.

In: Oil and Gas Development ...
Editor: Aaron Spearing

ISBN: 978-1-63321-779-9
© 2014 Nova Science Publishers, Inc.

Chapter 1

OIL AND GAS: UPDATED GUIDANCE, INCREASED COORDINATION, AND COMPREHENSIVE DATA COULD IMPROVE BLM'S MANAGEMENT AND OVERSIGHT[*]

United States Government Accountability Office

WHY GAO DID THIS STUDY

Federal and state regulatory agencies manage and oversee the development of federal and Indian oil and gas resources by issuing rules and guidance, reviewing drilling applications, and inspecting wells to ensure compliance with environmental, safety, and other regulations. In fiscal year 2013, federal and Indian energy production, including oil and gas, generated almost $15 billion in revenue. Recent advances in drilling technologies have greatly expanded the ability to develop oil and gas resources, particularly from shale and tight sandstone rock formations.

GAO was asked to review oversight of federal and Indian oil and gas resources. This report (1) discusses the changes to federal and selected state agencies' rules over the past 5 years and (2) examines the effectiveness of BLM's management and oversight of federal and Indian resources. GAO

---

[*] This is an edited, reformatted and augmented version of the United States Government Accountability Office publication, GAO- 14-238, dated May 2014.

reviewed federal and selected state agencies' rules and guidance, data from federal agencies, and other documentation. GAO selected a sample of 14 states based, in part, on their involvement with oil and gas development. This information cannot be generalized to all states.

WHAT GAO RECOMMENDS

GAO recommends that, among other things, BLM ensure that its rules governing oil and gas development are consistent with technological advances, improve coordination of inspections with states, and improve the timeliness of revenue sharing agreement reviews. In commenting on a draft of this report, BLM agreed with GAO's recommendations.

WHAT GAO FOUND

Federal and state agencies in states that GAO reviewed have taken or initiated some actions to change rules, in part, as a response to certain technological advances that have led to more than a 5-fold increase in annual production of domestic onshore oil and gas from shale and tight sandstone formations from 2007 through 2012 (see figure). For example, the Bureau of Land Management (BLM)—the key agency responsible for managing and overseeing oil and gas development on federal and Indian lands—proposed a new rule in 2012 to regulate hydraulic fracturing. In addition, in 2013, Texas updated its rules for well integrity by establishing new casing and cementing standards.

The effectiveness of BLM's management and oversight of federal and Indian oil and gas resources is hindered by a number of factors, including BLM's reliance on outdated rules and guidance, limited coordination with states, and delayed reviews of revenue sharing agreements. For example, BLM has not followed Interior's guidance to routinely review rules and update them consistently along with technological advances. As a result, some of BLM's rules and guidance governing oil and gas development have not kept pace with technological advancements, such as its guidance on well spacing, which, among other things, determines how to maximize oil and gas production from a formation. Improper spacing guidance could lead to lower levels of oil and gas production and, therefore, less revenue for the federal government and tribes. In addition, BLM has not developed formal agreements to improve coordination with state regulatory agencies, as called for in its internal guidance. As a result, BLM and state agency officials told GAO that agencies conduct duplicative inspections of some wells, while leaving other wells uninspected. Further, based on GAO's review of selected revenue-sharing agreements, BLM does not always review these agreements within the statutorily required time frames. These agreements identify the production allocation based on ownership and must be approved before operators can pay royalties to the resource owner. By statute, BLM is to review agreements within 120 days of receipt, but GAO found instances in which the agency had taken more than a year to review, resulting in delayed royalty payments to resource owners—who may rely on these payments for part or all of their income.

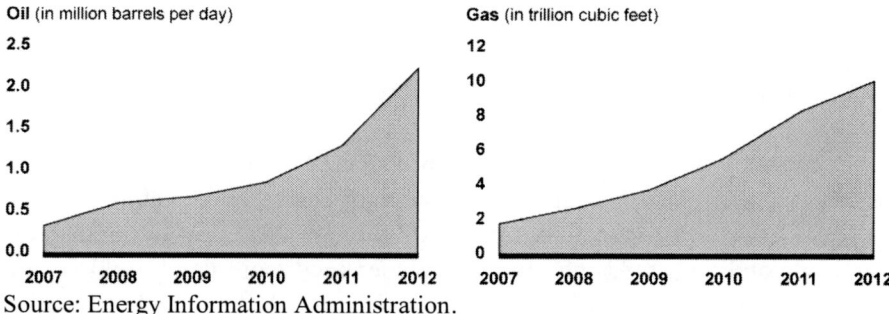

Source: Energy Information Administration.

Increased Domestic Oil and Gas Production from Shale and Tight Sandstone Formations from 2007 to 2012.

ABBREVIATIONS

AFMSS	Automated Fluid Minerals Support System
APD	application for permit to drill
BIA	Bureau of Indian Affairs
BLM	Bureau of Land Management
CAA	Clean Air Act
CWA	Clean Water Act
ELG	Effluent Limitations Guidelines and Standards
EPA	Environmental Protection Agency
FLPMA	Federal Land Policy and Management Act

FOGRMA	Federal Oil and Gas Royalty Management Act of 1982
FWS	Fish and Wildlife Service
GIS	geographic information system
IEED	Office of Indian Energy and Economic Development
NEPA	National Environmental Policy Act of 1969
NIOGEMS	National Indian Oil and Gas Evaluation and Management System
NPS	National Park Service
NTL	notice to lessees
SDWA	Safe Drinking Water Act
STRONGER	State Review of Oil and Natural Gas Environmental Regulations
USDA	Department of Agriculture

May 5, 2014

Congressional Requesters

In recent years, domestic onshore production of oil and gas has been steadily rising, providing an important source of energy and creating jobs for the United States. For example, from 2007 through 2012, annual production from shale and tight sandstone formations increased more than 6-fold for oil and approximately 5-fold for gas.[1] Horizontal drilling and hydraulic fracturing—technologies largely credited with spurring the boom in oil and gas production in the United States—have advanced significantly over the last decade and become increasingly common methods to recover oil and gas from geological formations once thought inaccessible, including shale and tight sandstone.[2] Oil and gas are produced on private, state-owned, federal, and Indian lands across the United States, and recovery of these resources from federal and Indian lands is a significant source of revenue for the federal government and tribes. For example, federal energy production generates one of the largest nontax sources of federal government revenue, accounting for more than $14 billion in fiscal year 2013, according to the Department of the Interior (Interior).[3] In addition, development of Indian energy resources provided almost $1 billion in revenue to tribes and individual Indians in fiscal year 2013.

Domestic onshore oil and gas development is governed by a framework of federal, state, tribal, and local laws and regulations. Several federal agencies—including Interior's Bureau of Land Management (BLM), Bureau of Indian

Affairs (BIA), and the Office of Natural Resource Revenue (ONRR), as well as the Environmental Protection Agency (EPA)—and state regulatory agencies have responsibility for oversight and management of oil and gas development on federal or Indian lands.[4] BLM has overseen the development of federal and Indian oil and gas resources for decades and is responsible for ensuring that these resources are developed in a timely, economically efficient, and environmentally sound manner. BLM's responsibilities include: (1) reviewing drilling plans and issuing permits for wells developing federal and Indian resources; (2) inspecting wells to ensure compliance with environmental, safety, and other regulations; (3) approving revenue- sharing agreements—known as communitization agreements—for federal and Indian resources; and (4) ensuring the maximum recovery of federal and Indian resources.[5]

Our past work has highlighted the importance of Interior's oversight of the roughly 700 million subsurface acres for which the federal government holds mineral rights, and we identified a range of weaknesses. For example, we reported, in March 2010, that Interior did not have reasonable assurance that it was collecting its share of revenue from oil and gas produced on federal lands, and that it continued to experience problems hiring, training, and retaining sufficient staff to provide oversight and management of oil and gas operations in part because of the department's human capital challenges.[6] In February 2011, we added Interior's management of federal oil and gas resources to our list of government programs at high risk of waste, fraud, abuse, and mismanagement or in need of broad reform.[7] In 2012, we reported that the extent and severity of environmental and public health risks associated with oil and gas development depend, in part, on federal and state regulations.[8]

In this context, you asked us to review the effectiveness of several aspects of federal oversight of oil and gas development on federal and Indian lands. Our objectives in this review were to (1) report on the actions, if any, that federal agencies and selected states have taken in the past 5 years to change rules or guidance in response to advances in oil and gas development technologies and (2) examine the effectiveness of BLM's management and oversight of federal and Indian resources.

To conduct this work, we reviewed federal and selected state laws, regulations, rules, and guidance regarding the management of domestic onshore federal and Indian oil and gas resources. To identify examples of changes to state rules and requirements governing oil and gas development over the past 5 years, we reviewed state laws, regulations, and rules and interviewed officials from 14 states we selected—Arkansas, California, Colorado, Louisiana, New Mexico, North Dakota, Ohio, Oklahoma,

Pennsylvania, South Dakota, Texas, Utah, West Virginia, and Wyoming. We selected these states based on a number of factors, including (1) experience levels with oil and gas development, (2) the presence of federal lands, and (3) the presence of Indian lands. Because this was a nonprobability sample, our results are not generalizable to all states but provide examples of state actions. We did not compare the changes in states' rules and regulations because development of oil and gas, as well as the environmental and public health risks, can vary based on the geological characteristics of the formation being developed. As a result, changes to rules in one state may not be appropriate for other states.

We obtained data on drilling inspections performed from fiscal year 2009 through fiscal year 2012 from BLM's Automated Fluid Minerals Support System (AFMSS)—the central database that BLM uses to track oil and gas information on federal and Indian lands. To assess the reliability of the inspection data, we performed electronic testing of the data and interviewed agency officials about the data. We determined that the inspection data were sufficiently reliable to present results on the number of inspections performed by BLM from fiscal year 2009 through fiscal year 2012. However, BLM data on wells and whether they were identified as high- or low- priority for inspections were not available for a significant number of wells; therefore we determined that these data were not sufficiently reliable to comprehensively assess BLM's compliance with its internal goal to inspect all high-priority wells. We also reviewed communitization agreements from BLM's Oklahoma office to identify dates that agreements were submitted to and approved by BLM. We selected BLM's Oklahoma office based on a number of factors, including: (1) the high levels of oil and gas development, (2) the probability that development will traverse multiple mineral owners, and (3) the availability of data.

To obtain additional perspectives on issues related to oil and gas rules and BLM's management and oversight, we interviewed a nonprobability sample of stakeholders representing numerous agencies and organizations, including officials from four agencies within Interior—BLM, BIA, the Fish and Wildlife Service (FWS), the National Park Service (NPS)—EPA, the U.S. Army Corps of Engineers (Corps), the Forest Service within the U.S. Department of Agriculture (USDA), several tribal nations, and representatives from the oil and gas industry. The federal offices were selected based on their regulatory oversight authorities or land management responsibilities.

We conducted this performance audit from September 2012 to May 2014 in accordance with generally accepted government auditing standards. Those standards require that we plan and perform the audit to obtain sufficient, appropriate evidence to provide a reasonable basis for our findings and conclusions based on our audit objectives. We believe that the evidence obtained provides a reasonable basis for our findings and conclusions based on our audit objectives.

BACKGROUND

This section includes an overview of (1) oil and gas resource ownership in the United States, (2) the oil and gas development process, and (3) key stakeholders and regulations.

Oil and Gas Resource Ownership in the United States

As the exploration and development of oil and gas has increased significantly in recent years, newly identified oil and gas reserves are creating opportunities across the United States. In some cases, oil and gas development occurs on lands where surface rights and the right to extract oil and gas resources—called mineral rights—are owned by different parties. In these cases, known as split estates, each party has its own set of property rights.[9] In a number of states, mineral rights are considered the dominant estate, meaning those rights take precedence over the rights of surface owners. However, the mineral owner is not free to completely disregard the rights of surface owners and generally must limit surface interference to what is reasonably necessary to extract oil and gas.

The severance of mineral and surface estates has occurred through various means, including the transfer of surface or mineral interests through private party contracts, federal land grants acts, homesteading laws, and other congressional actions. Specifically, federal land grant acts and homesteading laws created split estates by granting surface rights to homesteaders while reserving mineral rights for the federal government.[10] For instance, the Stock Raising Homestead Act of 1916 allowed a settler to claim 640 acres of land designated by the Secretary of the Interior as ranch land, but where the federal government retained mineral rights. Tribal and individual Indian estates were split through several congressional acts, including the General Allotment Act of 1887, the Indian Reorganization Act of 1934, and legislation to create or dissolve Indian reservations.[11] Split estates and changing ownership for both surface and mineral estates have created a patchwork of ownership patterns, and it is not uncommon to find federal, Indian, private, state, and county parcels of land intermingled together in some areas of the country—giving land ownership maps the appearance of a checkerboard.

Oil and Gas Development Process

Developing oil and gas resources involves a variety of activities and the specific steps to extract the resource vary based on a number of factors, such as the characteristics of the rock formation in which the resource is located. Our review focuses on the following activities: (1) planning and leasing; (2) permitting; (3) drilling, well construction, and hydraulic fracturing; and (4) production and unitization.

Planning and leasing. Operators locate suitable oil and gas targets using seismic methods of exploration.[12] Once a suitable target is found, the operator must acquire the necessary leases for the right to drill. For federal lands, the process begins when industry, the public, or BLM nominates a parcel of land for lease consideration. BLM then determines if the parcel is available for leasing based on a land use plan.[13] Consistent with a completed land use plan and associated environmental impact statement, BLM can offer for lease those resources identified in the plan. For Indian leases, BIA approves lease terms for parcels nominated by tribes or individual Indian mineral owners. State governments and private landowners also lease land to companies for the development of oil and gas resources.

Permitting. Operators that have obtained a lease for federal or Indian minerals must submit an application for permit to drill (APD) to BLM and obtain BLM's approval prior to drilling. A complete APD must include a drilling plan, among other things, which the agency uses to determine the technical adequacy of the proposed drilling operation. BLM reviews APDs and may approve them as submitted or approve them subject to certain conditions to ensure environmental protection, safety, or conservation of resources. After BLM approves an APD, an operator generally has 2 years to drill on leased lands before the APD expires.

Drilling, well construction, and hydraulic fracturing. Prior to drilling, an operator prepares the area of land where drilling will take place— referred to as a well pad.[14] In some cases, the operator will build new access roads to transport equipment to the well pad or install new pipelines to transport produced oil or gas. The operator may place storage tanks or construct pits on the well pad for temporarily storing fluids. Once the well pad is constructed, operators drill a vertical hole referred to as the wellbore. At several points in the drilling process, casing and cement may be inserted to maintain wellbore integrity and prevent any communication between the formation fluids and the wellbore.[15] After vertical drilling is complete, operators may extend the wellbore horizontally. Horizontal drilling is conducted by slowly angling the drill bit until it is drilling horizontally. Horizontal stretches of the well typically range from 2,000 to 6,000 feet long but can be as long as 12,000 feet. Throughout the drilling process, operators may release some natural gas, a byproduct of oil development, directly into the atmosphere—known as venting—or burn it—known as flaring—in response to maintenance needs, equipment failures, or because natural gas is a byproduct of oil development, and the area does not have the infrastructure to transport the resource to a processing facility.

Oil and natural gas are found in a variety of geologic formations, and the specific steps to extract the resource vary based on characteristics of the formation. For instance, conventional oil and natural gas are often found in deep, porous rock or reservoirs and flow under natural pressure to the wellbore and on to the surface after drilling. In contrast to the free-flowing resources found in conventional formations, the low permeability of some tight formations, including shale and tight sandstone, means that oil and gas trapped in the formation cannot move easily within the rock into the wellbore. In order to extract oil and gas from formations with low permeability, operators hydraulically fracture the formation. Hydraulically fracturing a well involves making small holes in the casing and cementing[16] and then injecting fracturing fluid and proppant—usually sand or ceramic beads—into the wellbore at high pressure to cause the rock in the target formation to fracture.[17] After the fractures are created, the proppant stays in the formation to hold open the fractures and allow the release of oil and gas into the wellbore and then up to the surface. Hydraulic fracturing has become an increasingly common method to recover oil and gas over the last decade across the country. For example, Interior estimates that about 90 percent of wells developing federal and Indian resources are stimulated using hydraulic fracturing. Some of the fracturing fluid that was injected into the well will return to the surface— commonly referred to as flowback—along with water that occurs naturally in the oil- or gas-bearing formation—collectively referred to as produced water. The produced water is brought to the surface and collected by the operator, where it can be stored on-site in impoundments, re-injected into underground wells drilled specifically for that purpose, transported to a wastewater treatment plant, or reused by the operator in other ways.

Production and unitization. Once a well is producing oil or natural gas, the operator may remove the equipment and temporary infrastructure associated with drilling and hydraulic fracturing, leaving only the infrastructure required to collect and process the oil or gas and produced water.

Regulations and rules that govern the location and density of wells—known as spacing rules—significantly affect oil and gas production. Spacing rules, established by state agencies and BLM, limit the number and location of wells that may be drilled into a reservoir and are necessary to (1) ensure the most efficient and economic recovery of oil and gas resources, (2) prevent the drilling of unnecessary wells, (3) minimize waste or unnecessary surface disturbance and the associated environmental effects, and (4) protect the correlative rights of mineral owners.

Spacing rules combine multiple tracts and leases into a single unified block that range in size—called a unit. For the purposes of oil and gas development, the tracts and leases included within the unit boundaries are considered a single entity. According to literature on oil and gas development, combining leases into a unit allows for wells to be located and drilled based on the reservoir or formation being developed, rather than lease boundaries, for the purpose of developing the resource efficiently and economically. Production from a well located within the unit is attributed to the entire area—meaning that all owners share a percentage of the drilling costs and royalties based on their percentage of surface ownership, regardless of where wells are located. The ideal size and shape of the unit may vary considerably depending, in part, on whether the geologic formation being developed is a conventional reservoir or a tight formation. For example, in a conventional oil deposit, in which oil flows relatively easily through the formation, ideal spacing may allow for fewer wells that are spaced further apart than wells in a tight formation, which may require wells and fractures to saturate the formation as the oil does not travel as easily through the formation. Figure 1 shows a hypothetical lease unit of multiple wells developing a conventional gas reservoir.

The advent of horizontal—or more broadly, directional—drilling may allow multiple wells to be drilled long distances from a single well pad so that the underlying oil and gas deposit is more uniformly accessed, while limiting the surface disturbance to a smaller area. Figure 2 shows a hypothetical unit consisting of multiple leases in a single unit in a shale formation.

Technological advances have also resulted in changes to the size and shape of units. Because horizontal drilling and a longer lateral wellbore allow an operator to access more of the resource, the ideal size of the unit may be larger than for a conventional formation, and the resource can be more efficiently developed with fewer wells and well pads, thereby minimizing surface disruption. If the unit is smaller than the length that the lateral wellbore can extend, then more wells and well pads than are ideal to efficiently develop the resource may be the result. Fewer well pads reduce disruption to the surface and surrounding area by, among other things, reducing the number of access roads and pipelines. In addition, units with horizontal wells are generally rectangular because tight geologic formations may require parallel wells to optimally recover the oil and gas.

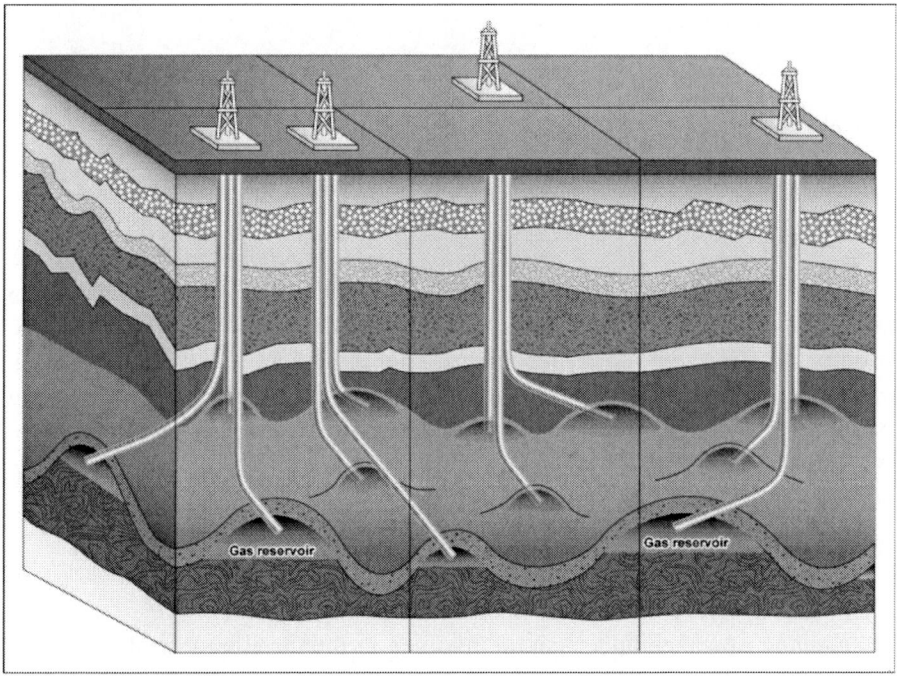

Source: GAO.

Figure 1. Example of a Multiple Lease Unit with Wells Developing a Conventional Gas Reservoir.

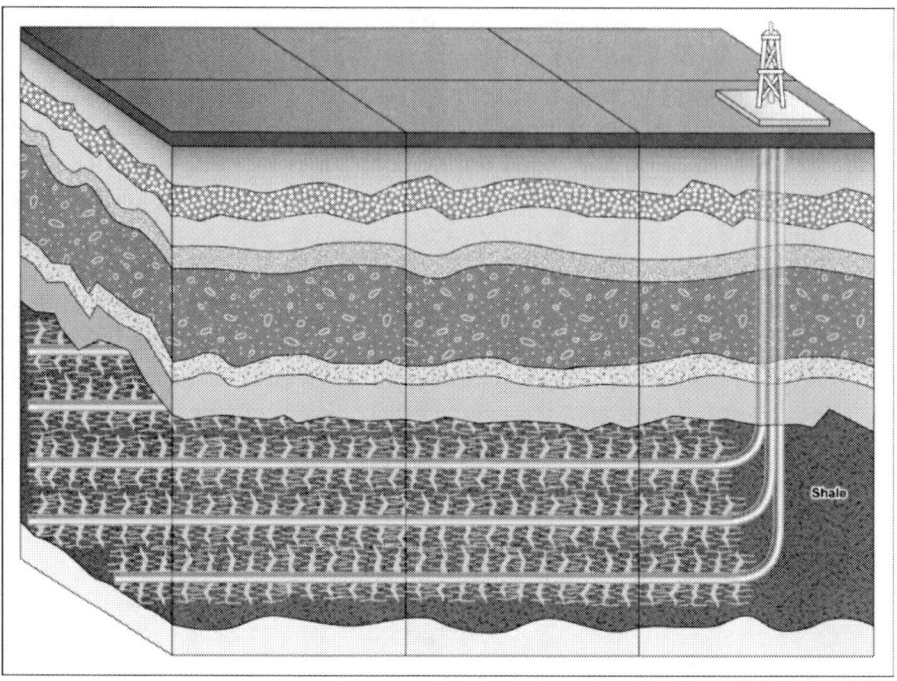

Source: GAO.

Figure 2. Example of a Multiple Lease Unit with Horizontal Wells Developing a Shale Formation.

If federal or Indian leases are included within the unit, the operator is required to obtain a communitization agreement approved by BLM for federal resources or BIA for Indian resources. The agreement identifies production allocations among mineral rights owners for the distribution of royalties and must be approved before operators can distribute royalty payments for federal and Indian leases.

Key Stakeholders and Regulations

A range of stakeholders can be involved in the development of federal or Indian resources, including federal and state agencies and tribes that administer key laws and regulations governing the various phases of oil and gas development.

Federal Agencies

BLM. Interior's BLM is responsible for most of the management and regulatory oversight of federal and Indian resources.[18] In addition to issuing leases, processing APDs, and reviewing communitization agreements, officials located in BLM field offices perform drilling inspections for wells that develop federal and Indian resources to ensure operations are in compliance with regulations.[19] BLM tracks data on federal and Indian oil and gas wells, including data on drilling inspections, in its AFMSS database. BLM guidance does not call for a drilling inspection of all wells, or specify when a well should be inspected. Instead, BLM guidance calls for drilling inspections on all wells rated high priority and allows for BLM to witness critical points in the drilling, blowout prevention, casing, and cementing processes. According to BLM officials, the criteria to identify high-priority wells is based on geologic factors, such as the need to protect usable water, high-pressure zones, or zones that contain hydrogen sulfide.

BLM is also responsible for protecting the federal government and Indian oil and gas owners from loss of royalties by ensuring that resources are developed in an economically efficient and environmentally sound manner.[20] To meet this responsibility, the agency must protect federal and Indian resources from being drained by adjacent wells and from mineral trespass—extracting minerals, including oil and gas, without the right to do so.[21] One way that the agency protects federal and Indian resources is by conducting drainage reviews. Drainage is the gradual removal of oil and gas from beneath a specific property by a producing well on an adjoining property. Drainage reviews evaluate wells on lands that are adjacent to or nearby federal and Indian leases to determine if drainage is occurring. BLM is also responsible for maintaining a system of records that identifies the location of federal and Indian resources.[22]

BIA. Interior's BIA administers oil and gas leasing for Indian resources, maintains current surface and subsurface ownership records, approves communitization agreements, and approves agreements for all surface related actions.[23] BIA also directs the Office of the Special Trustee to distribute royalty payments to the beneficial owners of Indian resources.

ONRR. Interior's ONRR manages revenues from federal and Indian resources. Revenue sources include royalties, rents, and bonuses generated by energy and mineral leases.

Office of Indian Energy and Economic Development (IEED). Interior's IEED serves tribes by assisting with the exploration, development, and management of their energy resources in order to generate new jobs and economic self-sufficiency.

Corps. Under Section 404 of the Clean Water Act (CWA), the Corps is authorized to issue or deny permits for the discharge of dredged or fill material into waters of the United States. For example, construction of a well pad or access road may require the discharge of dredged or fill material into waters of the United States. Under Section 10 of the Rivers and Harbors Act of 1899, the Corps has regulatory authority to oversee all work or structures in, over, or under navigable waters of the United States.[24] In connection with the Corps' mission, the agency has acquired certain private or easement lands that require written authorization of the Corps before oil and gas exploration and development may occur.

EPA. EPA administers and enforces key federal laws, such as the Safe Drinking Water Act (SDWA), the Clean Air Act (CAA), the CWA, and others, to protect human health and the environment. The SDWA is the main federal law that ensures the quality of drinking water. Two key aspects of the SDWA—the Underground Injection Control (UIC) program and the imminent and substantial endangerment provision—pertain to oil and gas development. Under its UIC program, EPA either implements or approves authority to states and eligible Indian tribes to implement a program regulating the injection of fluids underground, such as produced water from oil and gas development. EPA has lead implementation authority in 10 states and most Indian country; 33 states and two tribes have assumed primary responsibility, or primacy, for the UIC program; and 7 states share authority with EPA for the program.[25] The SDWA also gives EPA the authority to issue orders when the agency receives information about present or likely contamination of a public water system or an underground source of drinking water that may present an imminent and substantial endangerment to human health. EPA and its regional offices work with states and eligible Indian tribes that implement aspects of some of these laws, as well as additional state or tribal requirements. The agency's focus and obligations are to provide oversight, guidance and, where appropriate, rulemaking to achieve the best possible protection of air, water, and land.

Other land management agencies. Other land management agencies, including Interior's FWS and NPS, as well as the Forest Service, manage federal lands. FWS and NPS manage federal lands primarily for conservation and restoration of wildlife and their habitat and park resources, though some lands may be used for oil and gas development. The Forest Service manages land for multiple uses, such as recreation, timber harvesting, and energy development, including oil and gas.[26]

State Regulatory Agencies

In addition to their responsibilities for permitting and inspecting oil and gas development, states may implement and enforce certain aspects of federal requirements. For example, the CAA allows EPA to authorize states to issue air quality permits.[27] Under the CAA, EPA sets national ambient air quality standards for six criteria pollutants—ground level ozone, carbon monoxide, particulate matter, sulfur dioxide, nitrogen oxides, and lead—at levels it determines are necessary to protect public health and welfare. States then develop implementation plans establishing how they will attain air quality standards through regulation, permits, policies, and other means. Further, states may require operators to meet more stringent regulations than federal regulations.

Tribes

Indian tribes may have their own set of laws and regulations covering oil and gas development and may also implement and enforce various federal regulations. For example, under the Indian Mineral Development Act, subject to BIA approval, tribes may enter into an agreement to develop minerals in which the tribe owns an interest.[28] In addition, under the CAA, EPA's Tribal Authority Rule provides the opportunity for Indian tribes to seek EPA authorization to be treated in the same manner as a state government.[29] Under the rule, eligible tribes may submit a Tribal Implementation Plan for review by EPA. According to EPA documents, a Tribal Implementation Plan is a set of regulatory programs a tribe can develop and adopt to help attain and/or maintain national air quality standards.

FEDERAL AND SELECTED STATE AGENCIES HAVE TAKEN SOME ACTIONS TO CHANGE RULES IN RESPONSE TO TECHNOLOGICAL ADVANCES

Federal and state agencies in the states we reviewed have taken or initiated some actions to develop new rules and guidance or update existing rules in response to technological advances.

Federal Agencies

BLM, EPA, FWS, NPS, and the Corps have proposed or made changes to some rules and guidance governing oil and gas development in response to technological advances.

BLM. BLM has proposed a new rule to regulate hydraulic fracturing and is working to revise its existing onshore orders governing oil and gas development.

Hydraulic fracturing rule. In May 2012, BLM proposed a rule to regulate hydraulic fracturing for oil and gas development on federal and Indian lands to improve BLM's ability to manage well integrity and wastewater and require operators to publicly disclose chemicals present in fracturing fluid.[30] After initially receiving more than 177,000 public comments, the proposed rule was revised and proposed again in May 2013. The new proposal includes important safety standards and improves integration with existing state and tribal standards.[31] It also retains three components of the initial proposal by requiring operators to (1) disclose the chemicals they use in hydraulic fracturing fluid, (2) improve wellbore integrity to better ensure that fluids used during hydraulic fracturing are not contaminating groundwater, and (3) establish a water management plan for fluids that flow back to the surface. According to BLM officials, the agency has not established a date by which it expects the rule to be final.

Onshore orders. BLM is currently working to revise and codify some of its onshore orders and a notice to lessees (NTL) governing oil and gas measurement, site security, and production accountability.[32] According to BLM officials, the agency is currently updating and revising the various rules regulating production accountability and the measurement of oil and gas, which were last updated in 1989, to keep pace with changing industry practices and new technologies.[33] For example, BLM is proposing a new rule to establish standards to minimize the amount of venting and flaring at oil and gas production facilities on federal and Indian lands, as well as establishing standards for determining avoidable versus unavoidable losses of natural gas.

EPA. EPA has proposed or adopted several new rules to mitigate the environmental and safety risks associated with hydraulic fracturing, including the following.

Indian Country New Source Review rule. EPA finalized the Indian Country New Source Review rule, which outlines preconstruction permitting of air pollution control requirements for facilities, including future permits for oil and gas development, in Indian country under the authority of the CAA. More specifically, the Indian Country New Source Review rule is designed to create a pre-construction permitting mechanism for all minor sources in Indian Country and major sources in areas of Indian Country that do not meet the national ambient air quality standards. It imposes essentially the same permitting requirements as in areas of the country where air pollution levels persistently exceed the national ambient air quality standards in states that do not have approved state implementation plans. The Indian Country New Source Review rule was promulgated in July 2011 and will take full effect in 2014 for tribes that have not been approved by EPA to regulate air pollution under a tribal implementation plan.[34] According to EPA documentation, the agency plans to propose additional source categories for permits in the near future that will address additional oil and natural gas development activities.

UIC guidance. In February 2014, EPA published guidance on how its UIC permit writers should address hydraulic fracturing with diesel in the context of the UIC program.[35] The guidance outlines existing statutory and regulatory requirements for permitting wells where diesel fuels are used for hydraulic fracturing, technical recommendations for permitting those wells, and a description of diesel fuels for EPA UIC permitting.[36]

New Source Performance Standards. In August 2012, EPA published final New Source Performance Standards for the oil and natural gas sector.[37] According to EPA officials, this rule includes the first federal air standards for natural gas wells that are hydraulically fractured, along with requirements for several other sources of volatile organic compound (VOC) emissions that were not regulated at the federal level.[38] In addition to the VOC control requirements for gas well completions, the New Source Performance Standards require control of VOC emissions from storage vessels, compressors, pneumatic controllers and equipment leaks at gas processing plants. Although the New Source Performance Standards directly regulate VOC emissions, according to EPA officials, the control requirements also substantially reduce methane emissions.[39]

National Emission Standards for Hazardous Air Pollutants. In August 2012, concurrent with the New Source Performance Standards, EPA published final National Emission Standards for Hazardous Air Pollutants, updating its air toxics standards for oil and natural gas.[40] These standards cover hazardous air pollutants emitted from glycol dehydrators—used to remove water from gas—and storage vessels, and equipment leaks at natural gas processing plants.

Effluent Limitations Guidelines and Standards. EPA is updating its proposed rules for wastewater discharges related to oil and gas development. The CWA directs EPA to develop a plan for the issuance of new regulations, or revision of existing regulations, for technology-based treatment standards, known as Effluent Limitations Guidelines and Standards (ELG), for categories of industrial wastewater discharges. On a biennial cycle, EPA publishes a preliminary plan for public comment and then a final plan, identifying industrial categories for rulemaking. Based on its identification in the 2010 plan, EPA is currently developing a proposed rule to amend the ELGs for the Oil and Gas Extraction Category that is scheduled to be published in 2014. According to its preliminary plan published in August 2013, the agency is proposing to discontinue revisions to the Oil and Gas ELGs to regulate pollutant discharges from the development of gas from coalbed methane formations.[41] However, EPA officials told us the agency will continue to revise the Onshore Oil and Gas ELGs to provide additional controls, known as pretreatment standards, on wastewater pollutant discharges from the development of shale gas to publicly-owned wastewater treatment facilities.

FWS. FWS is drafting rules to consistently manage activities associated with non-federal oil and gas development, by private parties, on lands and waters of the National Wildlife Refuge System. While the federal government manages the surface lands in the system, in many cases, private parties own the mineral rights and have the legal right to develop underlying oil and gas resources. In August 2003, we reported that oversight and management of oil and gas activities varies widely among wildlife refuges and found that some refuges issue permits that establish operating conditions for oil and gas activities—giving these refuges greater control over oil and gas activities—while other refuges exercise little control.[42] According to FWS officials, the draft rule is intended to clarify and expand upon existing regulations that relate to the ownership of mineral rights in wildlife refuges.[43] On February 24, 2014, FWS published an Advance Notice of Proposed Rulemaking seeking comments to assist in the development of a proposed rule to avoid or minimize adverse effects on natural and cultural resources, wildlife- dependent recreation, public health and safety, and refuge infrastructure and management, while also ensuring a consistent and effective regulatory environment for oil and gas operators on refuge land. In the notice, FWS acknowledged that it is not proposing any specific approach for managing non-federal oil and gas operations.

NPS. NPS officials noted that the agency's current regulations governing oil and gas development on NPS lands have not been substantively updated for over 30 years. As a result, NPS is updating its regulations, referred to as "9B Regulations," which govern nonfederal oil and gas development in national parks.[44] According to NPS officials, about 60 percent of the 534 nonfederal oil and gas wells located on national parks fall outside the scope of the 9B Regulations or received grandfathered status because of regulatory exemptions, meaning that the wells are not required to obtain an NPS permit and meet NPS operating standards. Consequently, one NPS official said that significant park resources and values are not properly protected. According to NPS officials, under the revised regulations, no oil and gas wells will be exempt, and all wells must meet the same standards. NPS officials also stated that the agency analyzed its current regulations with regard to new proposals that use federal access and include horizontal drilling and hydraulic fracturing and found these regulations could fully address and mitigate activities for shale development. In addition, according to NPS officials, the revised regulations will clarify permit information requirements and operating standards associated with horizontal drilling and hydraulic fracturing, such as the public disclosure of chemicals used in hydraulic fracturing fluid. According to agency officials, NPS expects to issue the final regulations by 2015.

Corps. Corps officials told us the agency is developing national guidance for oil and gas development on Corps-managed lands, but the agency does not have an issuance date. Some Corps division and district offices recently issued guidance for oil and gas development. For example, the Corps' Southwestern Division developed guidance in 2011 for hydraulic fracturing and the removal of oil and natural gas from formations in close proximity to Corps dams and other key structures.[45] Officials from another Corps' district told us the office prepared interim guidance in 2012 to address the numerous requests from operators for permits to store water used for oil and gas development in Corps managed lakes.

State Agencies

All 14 of the states we reviewed have made some changes to rules governing oil and gas development in the past 5 years. At least 6 of the 14 states we reviewed that have changed or are proposing to change their oil and gas rules have done so as a result of regulatory reviews. Some of the oil and gas regulatory agencies in these states are required to review their oil and gas rules and regulations on a routine basis—once a year or up to every 5 years—

and amend regulations as needed. For example, North Dakota officials told us that they review and revise state rules at least once every 2 years to keep pace with advances in technology. Ohio officials told us they review and revise state rules every 5 years. In addition, 12 of the 14 states we reviewed participated in a review of their oil and gas rules and regulations by the State Review of Oil and Natural Gas Environmental Regulations (STRONGER).[46] STRONGER review teams have reviewed 22 state programs to identify, among other things, whether states have standards to prevent the contamination of groundwater and surface water.[47] In 2009, STRONGER formed a Hydraulic Fracturing Workgroup that examined hydraulic fracturing standards in six states—Arkansas, Colorado, Louisiana, Ohio, Oklahoma, and Pennsylvania. Officials we interviewed from at least 7 of the 14 states told us that the regular state reviews or recommendations made by STRONGER initiated positive changes to state rules and requirements governing oil and gas development. Specifically, we found that all of the states we reviewed changed some rules related to (1) the disclosure of chemicals used in hydraulic fracturing, (2) baseline testing, (3) well integrity, (4) water and waste management, (5) setback distances, (6) well spacing, and (7) air emissions, as described below.

Chemical Disclosure

The composition of fracturing fluid varies with the nature of the formation, but it contains about 98 percent water and proppant mixed with a small percentage of chemical additives, according to a report about shale gas development by the Ground Water Protection Council.[48] All of the states included in our review adopted or proposed rules requiring the public disclosure of certain information pertaining to the chemicals used in fracturing fluid (see figure 3).[49] For example, in 2012, Utah passed a rule regulating the hydraulic fracturing process, which, among other things, requires operators to report the chemicals used in fracturing fluids for each well within 60 days of the completion of the process.

Baseline Testing

In September 2012, we reported that oil and gas development poses a risk to water quality from contamination of surface water and groundwater as a result of spills and releases of produced water, chemicals, and drill cuttings; erosion from ground disturbances; or, underground migration of gases and chemicals.[50] During this review, we found some states recently established or amended rules requiring baseline water sampling prior to drilling to compare water conditions prior to and after drilling activities.[51] For example, Colorado established new rules in 2012 for statewide groundwater sampling and monitoring near new oil and gas wells. Colorado's revisions include initial baseline testing and subsequent monitoring of all available water sources, up to a maximum of four sources, within a half mile radius of a proposed oil and gas well, multi-well pad site, or dedicated injection well.

Well Integrity Requirements

In September 2012, we reported that well integrity is essential to isolate the well and protect the environment, including the groundwater. Some states we reviewed have revised regulations for well construction to require well-integrity testing, pressure monitoring testing, and new casing and cementing standards. For instance, in 2013, Texas updated its rules for cement quality, cementing, well equipment, centralizers, and well control. Specifically, the state established a minimum cement thickness for casing strings and consolidated requirements for well control and blowout preventers. The revised rule also adds requirements for wells that will be hydraulically fractured, such as pressure testing of the casing, and it requires operators to notify the state when a test fails and remedial work is necessary.

Water and Waste Management

We reported in September 2012 that oil and gas development poses a risk to water quality from spills or releases of toxic chemicals and waste, such as produced water. Some states that we reviewed are considering or have adopted changes to waste disposal and management rules that will address the recycling and transport of produced water, such as fluid storage options, pit liner requirements, and wastewater transportation. For example, in 2011, California established new rules to require secondary containment features around fluid containers, regular testing and maintenance of tanks and pipelines, and a detailed spill contingency plan. In another example, Oklahoma established new requirements in 2010 addressing the increased use of large flowback water recycling pits that pertain to the construction, siting, and maintenance of noncommercial storage pits and commercial recycling pits.

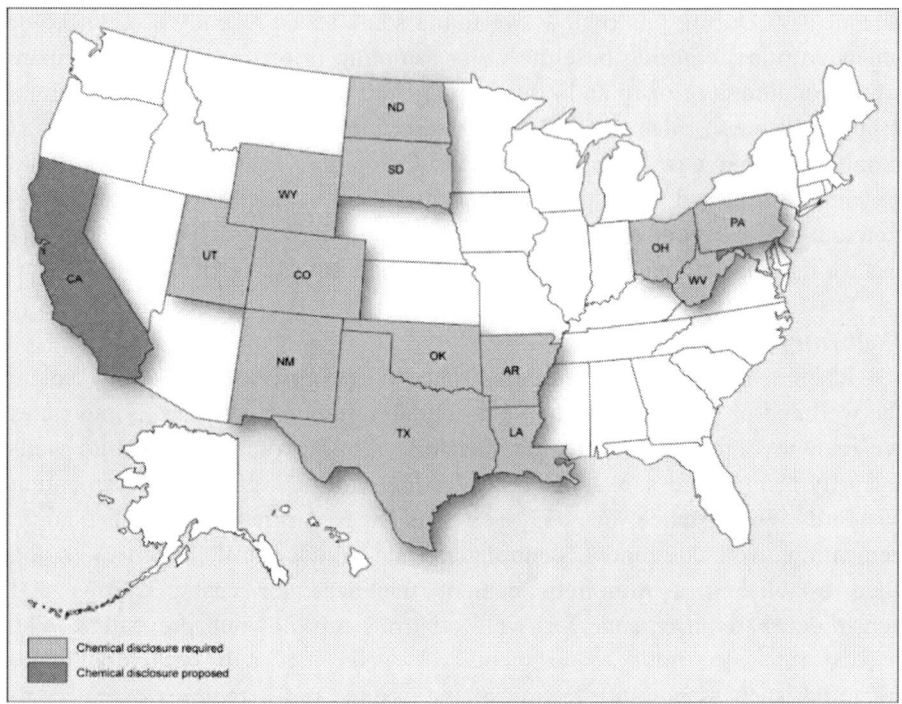

Source: GAO; Map Resources (map).

Figure 3. States with Proposed or Required Chemical Disclosure Rules That GAO Reviewed.

Setbacks

Setback rules, which regulate the distance between wells and other entities such as schools, homes, streams, and water wells, are created to protect the safety and welfare of the general public from environmental and nuisance effects resulting from oil and gas development, including spills, odors, noise, dust, and lighting. Some states we reviewed have recently amended their setback rules. For example, in 2013, Colorado established a uniform, state-wide 500-foot setback and imposed technically advanced best management practices and protective measures for all oil and gas locations within 1,000 feet of occupied buildings. The new rules stipulate that oil and gas locations may not be located within 1,000 feet of "high occupancy buildings," such as schools, day care centers, hospitals, nursing homes, and correctional facilities, without a public hearing and state approval.

Well Spacing

In response to the increased use of horizontal drilling, which allows operators to develop resources underlying larger land areas than vertical drilling, some states have implemented changes to requirements establishing spacing units—administratively combined contiguous leases—and the location of wells. For example, North Dakota changed its well spacing rules in 2012 to allow larger spacing units with more wells per unit. Specifically, the state modified existing well spacing rules to account for increases in horizontal drilling and long lateral wellbores. According to North Dakota officials, laterals are now extending from 4,000 to 9,000 feet long—allowing units to be from 640 acres to 1,280 acres. State officials told us that allowing larger units has doubled oil and gas recovery per unit while reducing surface disturbances caused by development. In another example, Oklahoma passed legislation in 2011 providing the state with the authority to modify or vacate existing drilling and well spacing units to accommodate horizontal drilling and long-lateral wellbore development that state officials say, in some cases, may exceed 5,280 feet in length.

Air Emissions

In September 2012, we reported that oil and gas development poses risks to air quality, generally as a result of engine exhaust from increased truck traffic, emissions from diesel-powered pumps used to power equipment, gas that is flared or vented for operational reasons, and unintentional emissions of pollutants from faulty equipment.[52] Officials from some states we reviewed told us that their state is considering rules to reduce air emissions. For example, Colorado is currently considering changes to a rule that would expand the state's existing air quality control program and establish new strategies to address the most significant sources of hydrocarbon emissions from oil and gas operators, such as requiring that the gas stream at well production facilities be connected to a pipeline or routed to a control device.

In addition to federal and state changes to rules governing oil and gas development, tribal governments are also taking actions to address changes in technological advances. For example, the Assiniboine and Sioux Tribes of the Fort Peck Reservation, Montana, developed best management practices for BIA and BLM to include in all leases and APDs as stipulations for approval. These best management practices are related to waste disposal, air emissions, noise and light pollution, surface disturbances, water use, and water protection.

For example, the tribes require operators to conduct baseline groundwater sampling prior to commencement of drilling and minimize truck usage on roads by using, when practical, alternate transport methods for liquids—such as pipeline and rail. The tribes also prohibit the use of pits to store produced water, drilling mud, and flowback fluids.

BLM'S MANAGEMENT AND OVERSIGHT OF FEDERAL AND INDIAN OIL AND GAS RESOURCES IS HINDERED BY SEVERAL FACTORS

The effectiveness of BLM's management and oversight of federal and Indian oil and gas resources is hindered by the following five factors: (1) outdated BLM rules and guidance, (2) limited formal coordination between BLM and state regulatory agencies on inspections, (3) incomplete BLM inspections and limited oversight of inspection activities, (4) incomplete data on onshore oil and gas resources and drilling activities, and (5) BLM delays reviewing communitization agreements.

BLM Does Not Ensure Its Rules and Guidance are Reviewed and Updated

BLM does not ensure that its rules and guidance governing oil and gas development have been reviewed and updated, as called for by Interior's guidance and executive order, to reflect technological advances, such as hydraulic fracturing and horizontal drilling. Without a documented process, most of BLM's rules governing oil and gas have not been reviewed and updated to reflect technological advances in over 2 decades.[53] In 1998, Interior released guidance requiring each bureau to evaluate its rules every 5 years to ensure that, among other things, they are needed and up to date. Moreover, in January 2011, Executive Order 13563 called for federal agencies to develop and submit a plan to facilitate the periodic review of existing significant regulations. The executive order states that the federal regulatory system should be based on the best available science. As such, the executive order requires agencies to consider how best to promote the retrospective analysis of rules that may be outmoded, ineffective, or excessively burdensome. In response, Interior outlined its plans in its 2011 Plan for Retrospective

Regulatory Review to ensure that its regulations are more functional, credible, and efficient by retrospectively reviewing them as part of its annual planning process.

However, according to BLM officials, some of its rules and guidance have not been periodically reviewed by the agency.[54] For example, BLM has not recently reviewed its guidance related to communitization agreements, which addresses well spacing. BLM's guidance states that the agency usually will not approve or, in the case of Indian lands, recommend approval of agreements combining more than 640 acres for oil or gas production irrespective of well location and federal or Indian acreage within a unit.[55] According to federal officials and industry representatives we spoke with, BLM's guidance on communitization agreements is inconsistent with industry practices because the limit of 640 acres does not accommodate horizontal drilling. As we previously mentioned, horizontal drilling allows operators to extend a well through a formation on a parallel plane, thereby exposing the well to more of the formation holding the oil and gas than a vertical well could achieve. As a result, a unit larger than 640 acres may be needed to allow operators to maximize oil and gas resource recovery and prevent waste. Due to the economic and strategic importance of oil and gas, BLM's guidance should provide reasonable assurance that well spacing will provide for maximum allowable production, thereby ensuring the federal government, tribes, and individual Indian resource owners are capitalizing on the development of the resource. Similarly, BLM officials told us that the agency has not updated its guidance on oil and gas drainage protection since 1999 and has not updated its guidance on mineral trespass since 2003.

BLM officials told us that the agency has conducted some reviews of rules and guidance, which are driven by program needs or as conditions warrant, but budgetary and staffing constraints have hindered regular comprehensive reviews of existing oil and gas rules and guidance. BLM officials also told us they are aware that some rules and guidance may be outdated and said that the agency will review and update guidance for communitization agreements and mineral trespass in 2014. Until BLM ensures its guidance for well spacing is up to date, the agency cannot ensure that federal and Indian resources are being managed and developed in a manner that provides for the maximum recovery potential of these resources, thereby maximizing the potential revenue provided to the federal government and tribes. It is also important to note that updating the guidance one time does not ensure compliance with Interior's guidance that calls for periodic reviews and updating. Currently, BLM does not have a documented process to ensure that oil and gas rules are periodically reviewed and updated. Without such a process, BLM cannot provide reasonable assurance that all oil and gas rules and guidance will be evaluated in the context of technological advances, as required by Interior's guidance and executive order.

Limited Formal Coordination between BLM and State Regulatory Agencies May Result in Duplicative Inspections

BLM and state regulatory agencies generally have overlapping authority to issue drilling permits and conduct drilling inspections for oil and gas development of federal and Indian resources, regardless of whether the surface estate owner is federal, Indian, state, or private. Specifically, BLM issues drilling permits to operators developing federal or Indian resources, regardless of the surface owner.[56] State regulatory agencies generally issue drilling permits to operators developing oil and gas resources within the boundary of the state, regardless of whether the owners of the resources or the surface are federal, state, tribal nations, individual Indians, or private entities.[57] Figure 4 shows a horizontal wellbore traversing mineral parcels owned by different entities. In this example, the operator is developing federal, Indian, state, and private oil and gas resources and would generally be required to obtain a drilling permit from both BLM and the state regulatory agency.

BLM and state regulatory agencies also have overlapping authorities to conduct drilling inspections. BLM has the authority to conduct a drilling inspection on any well that is being drilled into federal or Indian resources, regardless of surface ownership or if federal or Indian resources are collocated in a unit with private or state resources.

According to state regulatory officials we spoke with, state regulatory agencies generally have the authority to conduct drilling inspections of all operations within state boundaries. Both BLM and state regulatory agency officials told us that some wells have been inspected by both BLM and state regulatory inspectors, and other wells have not been inspected at all. In addition, state regulatory officials said that the increasing number of horizontal wells that will likely traverse federal or Indian resources and state or private resources is expected to increase the number of duplicative inspections.

BLM recognizes the importance of coordination, as evidenced by its internal guidance that directs BLM's state and field offices to pursue memorandums of understanding with state and tribal entities to help meet oil and gas inspection and enforcement goals. Moreover, according to BLM officials, the agency has limited staff to complete drilling inspections and therefore entering into agreements with the states could help leverage their limited resources and provide more effective oversight.[58] In some states, BLM has developed formal agreements through memorandums of understanding with state regulatory agencies— including in California, Colorado, Nevada, and Wyoming—that are intended to, in part, improve coordination of inspection activities and, in some cases, minimize duplication of oil and gas inspections. However, BLM has not developed formal agreements in other states with oil and gas development activities, such as New Mexico, North Dakota, Oklahoma, or Utah, and not all of the existing memorandums of understanding address coordination of inspections. Without formal agreements to coordinate inspections with state regulatory agencies or a process to leverage or utilize state inspection resources, BLM is not maximizing the number of oil and gas wells that can be inspected with existing resources. According to a state regulatory official, by minimizing or eliminating duplicate inspections between BLM and the state regulatory agency, the agencies could increase the total number of inspections conducted within the state. BLM officials told us the agency is working with the states of New Mexico and Utah to develop memorandums of understanding to improve coordination between the agencies regarding the management of energy development activities, including oil and gas inspections.

BLM Has Limited Assurance That Inspections Are Completed

BLM has limited assurance that inspections—the primary mechanism to ensure that operators are complying with various laws and regulations— are completed, as called for by its internal guidance, because the agency does not (1) inspect all high-priority wells, (2) have key data on wells, and (3) review and monitor inspection activities at its state and field offices.

BLM does not inspect all high-priority wells. According to BLM's Oil and Gas Inspection and Enforcement Handbook, the agency is to conduct drilling inspections for all wells identified as high priority. Our review of available data from AFMSS determined that more than 2,100 of the 3,702 wells that were identified as high priority in BLM's AFMSS database and drilled from fiscal year 2009 through fiscal year 2012 were not inspected. BLM officials told us that the agency has limited staff to complete drilling inspections, which is consistent with our prior report stating that Interior's human capital challenges have made it more difficult to carry out some oversight activities and that the agency conducted fewer inspections because of inspector vacancies.[59]

BLM does not have key data on wells. The extent to which BLM has conducted inspections of high-priority wells, as called for by its internal guidance, is not known because BLM's AFMSS database is missing key data. Specifically, of the more than 14,000 federal and Indian oil and gas wells drilled from fiscal year 2009 through fiscal year 2012 on lands managed by BLM that we reviewed, BLM's AFMSS database is missing data on whether 1,784 wells were identified as high- or low-priority.[60] Without these data, the extent to which the agency inspects high-priority wells is not known. BLM officials said that the agency is in the process of replacing AFMSS with a new system that, among other things, will require staff to indicate whether a well is high- or low-priority. As of December 2013, officials told us that the new system will be fully functional by fiscal year 2015.

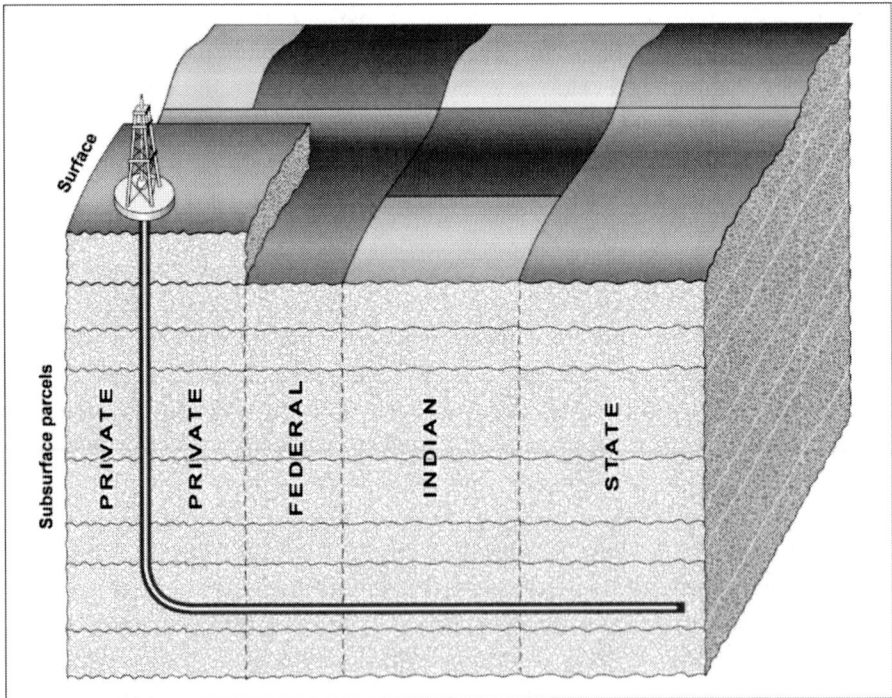

Source: GAO.

Figure 4. Example of a Horizontal Wellbore Traversing Mineral Parcels with Different Owners.

BLM does not review and monitor inspection activities at its state and field offices. BLM's headquarters office reported in fiscal year 2011 that the agency should establish annual reviews of its field offices to ensure oil and gas inspections are completed; however, BLM reviewed only one field office since issuing the report. Federal internal control standards call for agencies to compare actual performance with planned or expected results, monitor performance in the course of normal operations, and collect operational data to ensure that agencies are meeting their goals.[61] Without monitoring its field offices, BLM cannot provide reasonable assurance that they are meeting their inspection goals. BLM headquarters officials told us that they are developing a schedule to conduct field office reviews, but the agency does not have the resources needed to conduct the annual reviews. As we recently reported,[62] the dramatic increase in domestic oil and gas development has exacerbated BLM's long-standing challenges hiring and retaining staff needed to ensure effective oversight of oil and gas activities.

Data on Drilling Activities and Oil and Gas Resources Are Incomplete

BLM is also hindered in its ability to provide reasonable assurance that federal and Indian resources are properly managed and protected because not all BLM field offices have comprehensive and reliable data identifying the location of federal and Indian oil and gas resources or existing and new wellbores—data needed to protect oil and gas resources. In an internal guidance document, BLM states that efficient management of oil and gas resources requires key data—including the location of federal and Indian leased and unleased resources, the location of existing and new wellbores, and the status of existing wells. Several BLM offices we visited during our review have access to key data, including the location of federal and Indian resources and new and existing wellbores. For example, BLM's North Dakota field office uses a geographic information system (GIS) to compile and compare data on the location of federal and Indian resources with data identifying the location of new and existing wellbores to identify potential cases of drainage and mineral trespass. Officials from BLM's North Dakota field office told us that they used the GIS data to identify several potential cases of mineral trespass. In addition, Interior's IEED is working to identify and digitize the location of Indian oil and gas resources and make these data available through a GIS—the National Indian Oil and Gas Evaluation and Management System (NIOGEMS). A BIA official from the Fort Berthold office in North Dakota told us that NIOGEMS has helped officials identify potential mineral parcels for development and more effectively manage ongoing development.[63] However, not all BLM offices have access to these data. BLM officials from one field office recently reported that the number of wells extracting Indian resources without prior approval is a major concern for the agency. Without access to data on the location of resources and wells, BLM cannot provide reasonable assurance that federal and Indian resource owners' rights are protected.

Even in instances where key data are available, BLM's use of the data is inconsistent. For example, officials from one BLM field office told us that they have access to well data through a BLM contract with a third-party vendor but do not typically use the data because they do not have enough staff to review them. These findings are consistent with a BLM review conducted in August 2010 that found the agency had funded nationwide contracts to obtain information on the location of oil and gas wells for the purpose of identifying drainage cases but had not provided sufficient direction to staff responsible for using the data.[64] As a result, BLM does not have reasonable assurance that staff use the software and information to maximize the value of the contracts.

In addition, federal and tribal officials told us that the data on the location of federal and Indian resources are not reliable and that inaccurate boundary surveys for Indian resources are hindering development. This is a particular concern for riparian areas (narrow vegetated areas adjoining rivers, streams, and lakes) because riverbed boundaries are continuously changing, and lands must be resurveyed by BLM to establish legal ownership before oil and gas development can take place. According to BLM officials, the agency is aware of the concerns about the accuracy of data for oil and gas resources, particularly under riverbeds, but the agency is not able to complete additional surveys of federal or Indian resources due to funding shortfalls. Our findings are consistent with a report by Interior's Inspector General that raised concerns about the reliability and completeness of Interior's data on the location of Indian resources. The report stated that Interior's data on Indian resources was outdated; resulting in the potential loss of millions of dollars in revenue to tribes from oil and gas development.

BLM Does Not Always Approve Communitization Agreements within Required Time Frames

Based on our review of select communitization agreements, BLM does not always review communitizaton agreements within established time frames.[65] By statute, BLM is to issue all determinations of allocations of production for units within 120 days of a request for determination.[66] In addition, according to Interior's Interagency Standard Operating Procedures, BLM is to review communitization agreements for Indian leases within 30 days of a request.[67] However, we found examples where BLM has not issued allocation determinations or reviewed communitization agreements within these time frames. For instance, we reviewed data for 61 Indian and federal wells drilled from fiscal year 2009 through fiscal year 2012 in the state of Oklahoma and found that BLM averaged 229 days to approve Indian communitization agreements and 126 days to review federal communitization agreements. BLM officials told us that they are unable to review the agreements within the required time frame because, in part, the agency does not have the staff needed to review them. In 2012, an official with ONRR reported that the increase in oil and gas development in North Dakota resulted in a backlog of communitization agreements awaiting review. Without an approved communitization agreement, ONRR cannot distribute royalties correctly or in a timely manner because communitization agreements establish production allocation—information needed before operators can pay royalties.

As a result of these delays, approval of a communitization agreement may lag production and delay royalty payments to the federal government, tribal nations, and individual Indian oil and gas resource owners.[68] According to several tribal and federal officials we met with during our review, the delay to process communitization agreements has resulted in a delay of royalty payments. This is a concern because, according to a 2010 BLM report, individual Indian oil and gas resource owners may rely on revenue from oil and gas development to pay for daily expenses such as food, shelter, health, and education.[69]

BLM's Eastern Montana/Dakotas District Office has taken steps to improve its capacity to process communitization agreements within required time frames. Specifically, the BLM district office entered into a memorandum of agreement with the North Dakota Petroleum Council that, among other things, allows the council to fund additional staff at BLM to review communitization agreements. In addition, BLM officials told us the BLM district office began a pre-communitization agreement approval process to provide initial approval to communitization agreements as they are received from the operator. Through this approval, production allocations are established, and operators can begin to make royalty payments. After the initial approval, BLM conducts a thorough review of the communitization agreement to identify any concerns. In contrast, BLM officials in Oklahoma's field office told us that they do not believe this process would work for their office because it would add another step to the process after which BLM would still need to complete a full review at a later time. As a result, officials from Oklahoma's BLM office told us that they are not considering initial approval as an option for improving communitization agreement approval times. BLM headquarters officials told us that the agency is committed to allocating staff and resources to meet its needs by considering communitization agreements along with the needs of the entire oil and gas program, such as permitting and inspections. The agency stated it will develop a corrective action plan in fiscal year 2014 to address the pending communitization agreement applications.

CONCLUSION

As the use of technological advances in hydraulic fracturing and horizontal drilling have increased, federal and state regulatory agencies have undertaken some actions to evaluate and, in some cases, have proposed new rules or revised rules governing the development of federal and Indian resources, which can help to ensure a balance between protecting the environment, public health, and safety with the benefits of efficient production of domestic oil and gas on federal and Indian lands. However, BLM's continued reliance on outdated rules and guidance, limited coordination with state regulatory agencies, incomplete data on the location of resources and industry activities, and delayed reviews of communitization agreements hinder BLM's ability to effectively manage and oversee the development of federal and Indian resources. Specifically, BLM has not established a process to ensure that all rules and guidance are reviewed and periodically updated consistent with technological advances, as called for by Interior's guidance and executive order. Without such a process, BLM cannot provide reasonable assurance that oil and gas rules are keeping pace with and are consistent with technological advances. In addition, BLM's inspection program is fundamental to ensuring sound oil and gas operations. However, BLM may be missing opportunities to improve the efficiency and effectiveness of this program. In particular, BLM has not developed formal agreements, in the form of memorandums of understanding, as called for by its internal guidance, to coordinate inspections with state oil and gas regulatory agencies. As a result, it has conducted duplicative inspections of some wells and left other wells uninspected, as well as missing opportunities to deploy its inspection resources more effectively. Moreover, BLM does not have comprehensive data on the location of resources and new and existing wells. Even in instances where key data are available, BLM's use of the data is inconsistent because, in part, the agency has not provided sufficient direction to staff responsible for using the data. Without such data, the agency cannot accurately and efficiently identify whether federal and Indian resources are properly protected or that federal and Indian resources are at risk of being extracted without agency approval. Finally, because BLM is not able to review communitization agreements within required time frames, the federal government, tribes, and individual Indian oil and gas resource owners are not paid for the development of their resources in a timely manner.

RECOMMENDATIONS FOR EXECUTIVE ACTION

We recommend that the Secretary of the Interior direct the Director of the Bureau of Land Management to take the following four actions:

- Take steps to establish a documented process that will provide reasonable assurance that oil and gas rules and guidance is reviewed and periodically updated.
- Take steps to establish formal agreements with all relevant state regulatory agencies regarding oil and gas inspection activities; better leverage state inspection resources to avoid duplicative oversight activities, to the extent possible.
- Ensure that all BLM field offices with responsibility for protecting federal and Indian resources have the data and direction needed to efficiently identify the location of federal and Indian resources and new and existing oil and gas wells and consistently use the data to identify situations in which federal and Indian resources are at risk of being extracted without approval from BLM.

- Identify and take necessary steps to ensure communitization agreements are reviewed within required time frames.

AGENCY COMMENTS AND OUR EVALUATION

We provided a draft of this report for review and comment to Interior, EPA, the Department of Defense, and the Department of Agriculture. The four agencies provided technical comments on early or final drafts, which we incorporated as appropriate.

Interior and USDA also provided a letter in which they generally agreed with our findings and recommendations.

Frank Rusco
Director, Natural Resources and Environment

APPENDIX I. OBJECTIVES, SCOPE, AND METHODOLOGY

Our objectives for this review were to examine: (1) actions, if any, that federal agencies and selected states have taken to change rules governing oil and gas development over the past 5 years and (2) the effectiveness of the Bureau of Land Management's (BLM) management and oversight of federal and Indian resources.

To examine the actions, if any, that federal agencies and selected states have taken to change rules and requirements governing oil and gas development over the past 5 years, we reviewed federal and selected state laws, regulations, rules, and guidance regarding the management of domestic onshore oil and gas resources owned or held in trust by the federal government. For our review of federal laws, regulations, rules, and guidance, we focused on stakeholders involved in the development of federal and Indian resources, such as BLM, the Environmental Protection Agency (EPA), the U.S. Fish and Wildlife Service (FWS), the National Park Service (NPS), and the U.S. Army Corps of Engineers (Corps). In particular, we identified and reviewed several proposed or current federal laws governing oil and gas development, such as BLM's proposed hydraulic fracturing rule, BLM's onshore orders, BLM's current well spacing guidance, EPA's Indian Country New Source Review rule, EPA's draft guidance on hydraulic fracturing with diesel, and EPA's Effluent Limitations Guidelines and Standards for oil and gas.

In addition, we examined the extent to which BLM periodically reviewed oil and gas rules, as called for by the Department of the Interior's (Interior) guidance. We limited the evaluation of periodic updates to BLM because BLM has primary responsibility for oversight of oil and gas drilling operations on federal and Indian lands. For our review of state laws, regulations, and rules, we selected 14 states—Arkansas, California, Colorado, Louisiana, New Mexico, North Dakota, Ohio, Oklahoma, Pennsylvania, South Dakota, Texas, Utah, West Virginia, and Wyoming. We selected these states based on a number of factors, including: (1) experience levels with oil and gas development, (2) the presence of federal lands, and (3) the presence of Indian lands. We interviewed officials from all 14 states concerning any changes to rules and regulations governing oil and gas development in these states. We asked officials to identify changes in their state, such as changes to well spacing rules for oil and gas development. Because this was a nonprobability sample, our results are not generalizable to all states, but they provide examples of state actions. The development of oil and gas, as well as the environmental and public health risks, can vary based on the geological characteristics of the formation being developed. As a result, changes to rules in one state may not be appropriate for other states. Therefore, changes to regulations in one state cannot be compared to the activities in other states. Our review focused on the following oil and gas development activities: (1) planning and leasing; (2) permitting; (3) drilling, well construction, and hydraulic fracturing; and (4) production and unitization.

To examine the effectiveness of BLM's management and oversight of federal and Indian resources, we collected and assessed federal data; reviewed agency policy, guidance, and procedures; and, obtained additional perspectives from a variety of stakeholders. Specifically, we obtained data on drilling inspections performed from fiscal year 2009 through fiscal year 2012 from BLM's Automated Fluid Minerals Support System (AFMSS)—the central database that BLM uses to track oil and gas information on federal and Indian lands. To assess the reliability of the inspection data, we performed electronic testing of the data and interviewed agency officials about the data. We determined that the inspection data were sufficiently reliable to report the number of inspections performed by BLM from fiscal year 2009 through fiscal year 2012. However, we determined that BLM's data on whether a well was identified as high- or low-priority were not sufficiently reliable because data were not available for a significant number of wells. As such, we were not able to comprehensively assess BLM's compliance with the requirement to inspect all high-priority wells.

We also reviewed revenue sharing agreements—known as communitization agreements—from BLM's Oklahoma office to identify dates that agreements were submitted to and approved by BLM. The results of this analysis cannot be generalized to all BLM field offices. Rather, we selected Oklahoma based on a number of factors, including: (1) the high levels of oil and gas development, (2) the probability that development will traverse multiple mineral owners, and (3) the availability of data. Finally, to obtain additional perspectives on issues related to oil and gas rules and oversight, we interviewed a nonprobability sample of officials representing numerous agencies and organizations, including officials from BLM, BIA, FWS, NPS, EPA, the Corps, the Forest Service, several tribal nations, 14 state oil and gas regulatory agencies, and representatives from the oil and gas industry. The federal offices were selected based on their regulatory oversight authorities or land management responsibilities. We also conducted site visits to Colorado, North Dakota, Oklahoma, Texas, and Utah. We selected these locations based on a number of factors, including (1) experience level with oil and gas development, (2) states with federal lands, and (3) states with Indian lands.

We conducted this performance audit from September 2012 to May 2014 in accordance with generally accepted government auditing standards. Those standards require that we plan and perform the audit to obtain sufficient, appropriate evidence to provide a reasonable basis for our findings and conclusions based on our audit objectives. We believe that the evidence obtained provides a reasonable basis for our findings and conclusions based on our audit objectives.

End Notes

[1] Shale is a sedimentary rock that is predominantly composed of consolidated clay-sized particles.
[2] Hydraulic fracturing involves pumping water, sand, and chemical additives into oil and gas wells at high enough pressure to fracture underground rock formations and allow oil or gas to flow. When combined with horizontal drilling, hydraulic fracturing allows operators to fracture the rock formation along the entire horizontal portion of a well, increasing the number of pathways through which oil or gas can flow.
[3] The $14 billion in revenue includes energy production on both federal lands and offshore waters.
[4] Indian lands are defined to include federal property held in trust by the United States for individual Indians or Indian tribes; other designated lands held by individual Indians or Indian tribes; and public land owned by the United States that is designated for the sole use and benefit of individual Indians or Indian tribes. For a complete definition, see 20 U.S.C. §

7713(7). For the purposes of this report, Indian lands include both the surface and subsurface estates.

[5] For the purposes of this review, we use the term federal resources to include oil and gas minerals owned and managed by the federal government and Indian resources to include Indian oil and gas minerals that are held in trust by the United States for individual Indians or Indian tribes.

[6] GAO, *Oil and Gas Management: Interior's Oil and Gas Production Verification Efforts Do Not Provide Reasonable Assurance of Accurate Measurement of Production Volumes*, GAO-10-313 (Washington, D.C.: Mar. 15, 2010).

[7] GAO, *High-Risk Series: An Update*, GAO-11-278 (Washington, D.C.: February 2011).

[8] GAO, *Oil and Gas: Information on Shale Resources, Development, and Environmental and Public Health Risks*, GAO-12-732 (Washington, D.C.: Sept. 5, 2012).

[9] In some cases, the subsurface estate can be further segregated by mineral or stratum.

[10] These land grant statutes include the Coal Lands Acts of 1909 and 1910 (30 U.S.C. §§ 81, 83), the Stock Raising Homestead Act of 1916 (43 U.S.C. §§ 299, 301 (remainder of the act repealed by act of Oct. 21, 1976, Pub.L. No. 94-579, 90 Stat. 2787, 2792), and the Agricultural Entry Act of 1914 (30 U.S.C. §§ 121-123).

[11] Under the General Allotment Act, surface lands allotted to Indians were to remain in trust for 25 years and not allowed to be sold during this period. However, the time period could be extended by the BIA if deemed to be in the allottee's best interest. The trust period was extended indefinitely by the Indian Reorganization Act of 1934. As a result, allotted lands cannot be transferred or leased without approval of the BIA.

[12] The seismic method of exploration introduces energy into the subsurface through explosions in shallow "shot holes" by striking the ground forcefully (with a truck-mounted thumper), or by using vibrations. A portion of the energy returns to the surface after being reflected from the subsurface strata. This energy is detected by surface instruments, called geophones, and the information carried by the energy is processed by computers to interpret subsurface conditions.

[13] Land use plans identify federal lands and mineral resources available for oil and gas development and other activities and guide management decisions and actions on public lands. As part of developing or revising land use plans, BLM is required under the National Environmental Policy Act of 1969 (NEPA), as amended, to evaluate likely environmental effects of decisions in the plan, such as selecting areas for oil and gas development. The Federal Land Policy and Management Act (FLPMA) requires the Secretary of the Interior to evaluate these plans for potential revision. In addition, Interior generally prepares an environmental impact statement—a detailed statement of the likely environmental effects of the proposed action—in preparing land use plans.

[14] In some cases, multiple wells will be located on a single well pad.

[15] Casing is a metal pipe that is inserted inside the wellbore to prevent high-pressure fluids outside the formation from entering the well and to prevent drilling mud inside the well from fracturing fragile sections of the wellbore.

[16] A perforating tool is inserted in the casing in order to make the holes that allow the fracturing fluid to enter into the formation.

[17] Fracturing fluid generally contains a number of chemical additives, each of which is designed to serve a particular purpose. For example, operators may use a friction reducer to minimize friction between the fluid and the pipe, acid to help dissolve minerals and initiate cracks in the rock, and a biocide to eliminate bacteria in the water that cause corrosion. The number of chemicals used and their concentrations depend on the particular conditions of the well.

[18] BLM's Onshore Oil and Gas Orders and notice to lessees (NTL) implement and supplement the oil and gas regulations for conducting oil and gas operations on federal and Indian lands (43 C.F.R. § 3164.1).

[19] BLM has 33 offices with oil and gas responsibilities located primarily in the Mountain West. BLM's headquarters, state, and district offices oversee and provide guidance and support to

the field offices that implement BLM's oil and gas program. BLM's authority for inspecting wells is, in large part, derived from the Federal Oil and Gas Royalty Management Act of 1982 (FOGRMA). The act requires, among other things, the Secretary of the Interior to develop guidelines that specify the coverage and frequency of inspections.

[20] 25 C.F.R. §§ 211-212, 43 C.F.R. § 3100, and BLM Instruction Memorandum No. 99-051, Jan. 27, 1999.

[21] Mineral trespass can be done intentionally (with knowledge that it has crossed subsurface property lines) or inadvertently (under a genuinely mistaken belief of a right to extract the minerals). Secretarial Order 3215 provides that management of Interior's Indian trust assets (BLM and BIA) must protect Indian trust assets from loss, damage, unlawful alienation, waste, and depletion, and maintain a system of records that identifies the location and value of Indian resources.

[22] 43 C.F.R. § 3160.0-9.

[23] BIA's decisions and actions are to be based on the best interest of the Indian mineral owner.

[24] Pub. L. No. 69-560, 44 Stat. 1010; Pub. L. No. 71-520, 46 Stat. 918.

[25] EPA uses the term Indian country, defined in 40 C.F.R. § 71.2, to include all land within the limits of an Indian reservation under the jurisdiction of the U.S. government, all dependent Indian communities within the borders of the United States, and all Indian allotments, the Indian titles to which have not been extinguished.

[26] In order for oil and gas development to occur within a national park, the NPS Regional Director must consent to the lease and permit, and can do so only upon determination that the activity permitted will not have significant adverse effect upon the resources or administration of the unit. 43 C.F.R. § 3109.2(b).

[27] In September 2012, we reported that oil and gas development poses risks to air quality, generally as the result of (1) engine exhaust from increased truck traffic, (2) emissions from diesel-powered pumps used to power equipment, (3) gas that is flared (burned) or vented (released directly into the atmosphere) for operational reasons, and (4) unintentional emissions of pollutants from faulty equipment or impoundments—temporary storage areas.

[28] Pub. L. No. 97–382, Dec. 22, 1982, 96 Stat. 1938 (codified at 25 U.S.C. §§ 2101-2108).

[29] According to EPA's website, 32 tribes have received eligibility determinations under the Tribal Authority Rule; 2 tribes have been approved to implement their own air quality control plans to address air quality issues on their reservations, with several more under development; and 1 tribe has received a delegation to implement a Title V operating permit program for its reservation.

[30] 77 Fed. Reg. 27691.

[31] 78 Fed. Reg. 31636.

[32] Several of these orders are being revised in response to recommendations from GAO, Interior's Office of Inspector General, and Interior's Subcommittee on Royalty Management.

[33] The revisions are being considered for onshore order #3, which applies to site security; onshore order #4, which applies to the measurement of oil; onshore order #5, which applies to the measurement of gas; and NTL-4A, which applies to waste prevention and the beneficial use of oil and gas.

[34] On January 17, 2014, the United States Court of Appeals for the District of Columbia Circuit vacated EPA's Indian Country New Source Review Rule with respect to non- reservation lands. In *Oklahoma Dept. of Env. Quality v. EPA, No. 11-1307 (D.C. Cir. Jan. 17, 2014)*, the court found that EPA overstepped its statutory limitations by applying a Federal Implementation Plan governing air permitting to lands outside formal reservation boundaries without first demonstrating that a tribe has jurisdiction over such lands.

[35] The UIC program was not historically used to regulate hydraulic fracturing, even though fracturing involves the injection of fluid underground. However, in 2001, a U.S. Court of Appeals found that EPA's UIC requirements apply to hydraulic fracturing (*Legal Envtl. Assistance Found., Inc. v. EPA*, 276 F. 3d 1253 (11th Cir. 2001) *reh'g en banc denied*, 34 Fed. App'x 392 (11th Cir. 2002), *cert. denied*, 527 U.S. 989 (2002). In 2005, Congress

amended the SDWA to exempt hydraulic fracturing from the UIC program, except if diesel fuel is injected as part of hydraulic fracturing. Therefore, the SDWA, as amended, continues to require the regulation of hydraulic fracturing using diesel.

[36] We are conducting a separate review of EPA's UIC program.

[37] U.S. EPA, Oil and Natural Gas Sector: New Source Performance Standards and National Emission Standards for Hazardous Air Pollutants Reviews Final Rule, 77 Fed. Reg. 49490 (Aug. 16, 2012) (codified at 40 C.F.R. Parts 60 and 63).

[38] According to EPA's website, VOCs are emitted as gases from certain solids or liquids. VOCs include a variety of chemicals, some of which may have short- and long-term adverse health effects.

[39] Methane emissions represent a waste of resources and a fractional contribution to greenhouse gas levels.

[40] 77 Fed. Reg. 49490 (Aug. 16, 2012) (codified at 40 C.F.R. Parts 60 and 63).

[41] To develop natural gas from coalbed formations, water from the coal bed is withdrawn to lower the reservoir pressure and allow the methane to release from the coal.

[42] GAO, *National Wildlife Refuges: Opportunities to Improve the Management and Oversight of Oil and Gas Activities on Federal Lands,* GAO-03-517 (Washington, D.C.: Aug. 28, 2003).

[43] 50 C.F.R. § 29.32.

[44] 36 C.F.R. Pt. 9, Sbpt. B.

[45] Corps guidance recommends establishing a 3,000 foot exclusionary zone around critical structures, including dams, spillways, intake structures, and embankments. No oil and gas exploration and production would be allowed in exclusionary zones.

[46] STRONGER is a non-profit, multi-stakeholder organization formed in 1999 to help states share innovative techniques, environmental protection strategies, and opportunities to improve environmental regulation and oversight of the exploration, development and production of oil and natural gas.

[47] The STRONGER review teams are composed of stakeholders from the oil and gas industry, state environmental regulatory programs, and members of environmental and public interest communities.

[48] Ground Water Protection Council and ALL Consulting. "Modern Shale Gas Development in the United States: A Primer." Prepared for the Department of Energy and National Energy Technology Laboratory (April 2009).

[49] California SB4 was enacted in September 2013, with regulations to be adopted for provisions governing, among other things, disclosure of the composition of hydraulic fracturing fluid by January 1, 2015.

[50] GAO-12-732.

[51] In some states, operators routinely conduct baseline testing. For example, we reported that, although Pennsylvania does not require baseline testing, the state law presumes operators to be liable for any pollution of water wells within 2,500 feet of a shale oil or gas well that occurs within 12 months of drilling activities, therefore, many operators in the state routinely conduct their own baseline testing of nearby water wells prior to drilling.

[52] GAO-12-732.

[53] Six of the seven Onshore Oil and Gas Orders were promulgated in the late 1980s and early 1990s. Onshore Order 1 was revised more recently in 2007, and it pertains to the approval of operations on federal and Indian oil and gas leases.

[54] BLM officials told us that they consider updating guidance to be a part of the process to review rules.

[55] BLM Communitization Manual 3160-9.

[56] According to BLM officials we spoke with, the agency would not issue a drilling permit if an operator was cutting through federal or Indian minerals, but not developing from those minerals. In commenting on a draft of this report, BLM stated that there is a difference of opinion among BLM officials on whether BLM would issue a drilling permit in this scenario.

[57] In addition to drilling permits, operators must obtain other types of permits from various federal, state, tribal, and local agencies to develop oil and gas resources, depending on a number of factors, including location of the well and the operator's actions. For example, the Wyoming Department of Environmental Quality requires permit coverage for stormwater discharges from all construction activities disturbing 1 or more acres.

[58] We recently reported that Interior's human capital challenges have made it more difficult to carry out some oversight activities and that the agency conducted fewer inspections because of inspector vacancies. GAO, *Oil and Gas: Interior Has Begin to Address Hiring and Retention Challenges but Needs to Do More*, GAO-14-205 (Washington, D.C.: Jan. 31, 2014).

[59] GAO-14-205.

[60] Our findings from this review are consistent with those of our August 2013 report—see GAO, *Oil and Gas Development: BLM Needs Better Data to Track Permit Processing Times and Prioritize Inspections*, GAO-13-572 (Washington, D.C.: Aug 23, 2013)—and those of Interior's Office of Inspector General. In August 2013, we found that BLM's AFMSS does not include information needed to identify wells that pose the greatest environmental risk—key criteria needed for determining which wells will receive environmental inspections. In addition, in December 2010, Interior's Office of Inspector General also reported that BLM cannot completely verify which oil and gas wells have or have not been inspected because of concerns about the reliability of data in the AFMSS database. Specifically, the report states that data entry controls are inadequate.

[61] GAO, *Internal Control: Standards for Internal Control in the Federal Government*, GAO/AIMD-00-21.3.1 (Washington, D.C. November 1999).

[62] GAO-14-205.

[63] In 2012, Interior's Inspector General reported that NIOGEMS represents a significant improvement over Interior's current database for managing oil and gas activities on Indian lands.

[64] BLM, *Oil and Gas Drainage and Indian Diligent Development: Operations and Policy Review* (August, 2010).

[65] The Secretary of the Interior receives requests for approval of allocation schedules pursuant to federal communitization agreements. Additionally, according to BLM officials, the agency is to review communitization agreements to ensure that, among other things, they conform to spacing rules and are in the best interest of the United States or the Indian mineral owner.

[66] 30 U.S.C. § 1721(l).

[67] BLM officials told us that the memorandum of agreement that calls for the review of communitization agreements within 30 days was not official because it was not signed. However, several BLM officials we met with told us that they operated under the agreement as if it was official.

[68] Our findings are consistent with a 2006 Royalty Policy Committee report that recommended BLM review annually the status of communitization agreements awaiting field office approval and communitization agreement approval timelines to identify any prioritization, resource allocation, and/or training needs. The Royalty Policy Committee is an advisory committee appointed by the Secretary of the Interior that reviews and provides advice to Interior on mineral-related issues and policies.

[69] BLM, *Oil and Gas Drainage and Indian Diligent Development: Operations and Policy Review* (August, 2010).

In: Oil and Gas Development ...
Editor: Aaron Spearing

ISBN: 978-1-63321-779-9
© 2014 Nova Science Publishers, Inc.

Chapter 2

U.S. CRUDE OIL AND NATURAL GAS PRODUCTION IN FEDERAL AND NON-FEDERAL AREAS[*]

Marc Humphries

SUMMARY

In 2013, the price of oil averaged $98 per barrel (West Texas Intermediate spot price), up from $94 per barrel in 2012. Prices remain high in early 2014 (near $100 per barrel) and are projected by the Energy Information Administration (EIA) to average in the mid-$90 per barrel range through 2014. A number of proposals designed to increase domestic energy supply, enhance security, and/or amend the requirements of environmental statutes are before the 113th Congress. A key question in this discussion is how much oil and gas is produced in the United States each year and how much of that comes from federal versus non-federal areas. Oil production has fluctuated on federal lands over the past five fiscal years but has increased dramatically on nonfederal lands. Non-federal crude oil production has been rapidly increasing in the past few years partly due to favorable geology and the relative ease of leasing from private parties, rising by 2.1 million barrels per day (mbd) between

---

[*] This is an edited, reformatted and augmented version of a Congressional Research Service publication R42432, prepared for Members and Committees of Congress, dated April 10, 2014.

FY2009-FY2013, causing the federal share of total U.S. crude oil production to fall by nearly 11%.

Natural gas prices, on the other hand, have remained low for the past several years, allowing gas to become much more competitive with coal for power generation. The shale gas boom has resulted in rising supplies of natural gas. Overall, annual U.S. natural gas production rose by about four trillion cubic feet (tcf) or 19% since FY2009, while production on federal lands (onshore and offshore) fell by about 28%. Natural gas production on non-federal lands grew by 33% over the same time period. The big shale gas plays are primarily on non-federal lands and are attracting a significant portion of investment for natural gas development.

The number of producing acres may or may not be a function of how many acres are leased, and the number of acres leased may or may not correlate to the amount of production, but in recent years, some members of Congress have proposed a $4/acre lease fee for non-producing leases. This proposal grew out of the efforts to open more public land and water (offshore) for oil and gas drilling and development when gasoline prices spiked in 2006-2008. Some in Congress noted that there were many leases they believed were not being developed in a timely fashion, while at the same time, others in Congress were pushing for greater access to areas off-limits (such as the Arctic National Wildlife Refuge (ANWR) and areas under leasing moratoria offshore). Higher rents for offshore leases were imposed by the Secretary of the Interior in 2009 to discourage holding unused leases and to move more leases into production, if possible.

Another major issue that Congress may seek to address is streamlining the processing of applications for permits to drill (APDs). Some members contend that this would be one way to help boost energy production on federal lands. After a lease has been obtained, either competitively or non-competitively, an APD must be approved for each oil and gas well. Despite the new timeline for review (under the Energy Policy Act of 2005, P.L. 109-58), it took an average of 307 days for all parties to process (approve or deny) an APD in 2011, up from an average of 218 days in 2006. The difference, however, is that in 2006 it took the Bureau of Land Management (BLM) an average of 127 days to process an APD, while in 2011 it took BLM 71 days. In 2006, the industry took an average of 91 days to complete an APD, but in 2011, industry took 236 days. The BLM stated in its FY2012 and FY2013 budget justifications that overall processing times per APD have increased because of the complexity of the process.

INTRODUCTION[1]

In 2013, the price of oil averaged $98 per barrel (West Texas Intermediate spot price), up from $94 per barrel in 2012. Prices remain high in early 2014 (near $100 per barrel) and are projected by the Energy Information Administration (EIA) to average in the mid-$90 per barrel range through 2014. A number of proposals designed to increase domestic energy supply, enhance security, and/or amend the requirements of environmental statutes are before the 113th Congress. A key question in this discussion is how much oil and gas is produced in the United States each year and how much of that comes from federal versus non-federal areas. Oil production has fluctuated on federal lands over the past five fiscal years but has increased dramatically on nonfederal lands. Non-federal crude oil production has been rapidly increasing in the past few years, partly due to favorable geology and the ease of leasing, rising by 2.1 million barrels per day (mbd) between FY2009 and FY2013, causing the federal share of total U.S. crude oil production to fall by nearly 11%.

Natural gas prices, on the other hand, have remained low for the past several years, allowing gas to become much more competitive with coal for power generation. The shale gas boom has resulted in rising supplies of natural gas. Overall, annual U.S. natural gas production rose by about four trillion cubic feet (tcf) or 19% since FY2009, while production on federal lands (onshore and offshore) fell by about 28%. Natural gas production on non-federal lands grew by 33% over the same time period (see *Table 2*). The big shale gas plays are primarily on nonfederal lands and are attracting a significant portion of investment for natural gas development.

This report examines U.S. oil and natural gas production data for federal and non-federal areas with an emphasis on the past five years of production.[2]

U.S. CRUDE OIL PRODUCTION: FEDERAL AND NON-FEDERAL AREAS (FISCAL YEAR)

Historically, according to Department of the Interior (DOI) data, crude oil production on federal lands was consistently under 20% of total U.S. production until the late 1990s. Annual production then surged on federal lands (primarily offshore), rising to over 30% in the early 2000s and reaching a high point of about 36% in FY2010.[3] As a result of recent production increases on non-federal lands, the question is raised whether non-federal lands might regain a more dominant position of roughly 80%-85% of total U.S. crude oil production. The fact remains, however, that there are 5.3 billion barrels of proved oil reserves located on federal acreage onshore and another 5.6 billion barrels of proved reserves offshore (nearly all in the Gulf of Mexico). Taken together, U.S. federal oil reserves equal about 43% of all U.S. crude oil reserves, which are estimated at 29 billion barrels, according to the EIA.[4] Proved oil reserves are amounts accessible under current policy, prices, and technology.

Crude oil production on federal lands, particularly offshore, is likely to continue to make a significant contribution to the U.S energy supply picture and could remain consistently higher than previous decades, but it could still fall as a percent of total U.S. production, if production on non-federal lands continues to rise at a faster rate.

Table 1. U.S. Crude Oil Production: Federal and Non-Federal Areas FY2009-FY2013

Fiscal Year	U.S. Total	Non-Federal	(Barrels per day)	Federal Offshore	Federal Onshore
			Total Federal (% of U.S. Total)		
2013	7,235,000	5,576,700	1,658,300	1,294,000	364,465
			(23)		
2012	6,241,000	4,598,000	1,643,000	1,303,300	339,700
			(26.3)		
2011	5,552,000	3,826,500	1,725,500	1,415,600	309,900
			(31)		
2010	5,438,800	3,463,700	1,975,100	1,680,300	294,800
			(36.3)		
2009	5,233,000	3,464,400	1,768,600	1,482,900	285,700
			(33.8)		

Source: Federal data obtained from the Office of Natural Resources Revenue (ONRR) Statistics, as of February 2014, http://www.onrr.gov (using sales year data), March 2014.

Notes: U.S. Fiscal Year Total data derived from EIA monthly production data contained in its publication Petroleum and Other Liquids, U.S. Field Production of U.S. Crude Oil, March 28, 2014, http://www.eia.gov. Data includes lease condensate, defined by EIA as a liquid hydrocarbon recovered from lease separators or field facilities at associated and non-associated natural gas wells.

There is however, continued interest among some in Congress to open more federal lands for oil and gas development (e.g., the Arctic National Wildlife Refuge (ANWR) and areas offshore) and increase the speed of the permitting process. But having more lands accessible may not translate into higher levels of production on federal lands, as industry seeks out the most promising prospects and higher returns on more accessible non-federal lands.

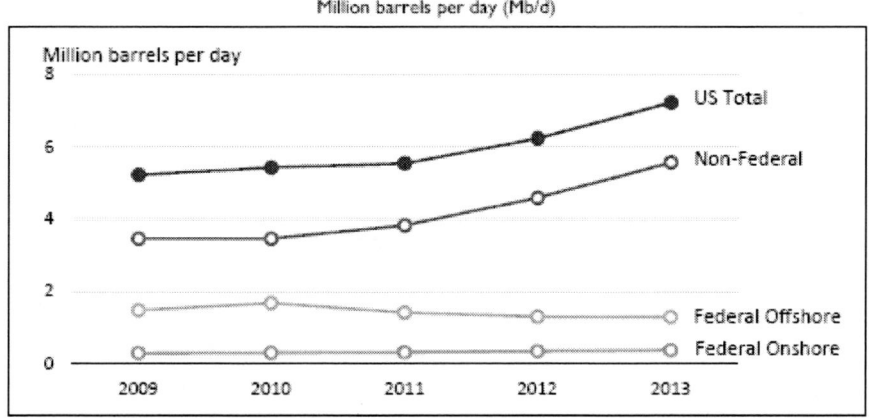

Source: Federal data obtained from ONRR Statistics, http://www.onrr.gov (using sales year data). Figure created by CRS.

Figure 1. U.S. Crude Oil Production: Federal and Non-Federal Areas, FY2009-2013.

U.S. NATURAL GAS PRODUCTION: FEDERAL AND NON-FEDERAL AREAS (FISCAL YEAR)

Natural gas production in the United States overall has dramatically increased each year since 2009, while production on federal lands has declined each year over the same period. Much of the decline can be attributed to offshore production falling by about 50%. Onshore production declines were less dramatic. Federal natural gas production has fluctuated from around 30% of total U.S. production for much of the 1980s through the early 2000s (34% of U.S. total in 2003), after which there began a steady decline through 2013.[5] This picture of natural gas production is much different than that of federal crude oil in that federal natural gas had accounted for a much larger portion of total U.S. natural gas over that past few decades.

Any increase in production of natural gas on federal lands is likely to be easily outpaced by increases on non-federal lands, particularly because shale plays are primarily situated on nonfederal lands and are where most of the growth in production is projected to occur.

U.S. dry gas proved reserves are estimated at about 334 tcf by the EIA,[6] of which the federal share is about 25% (69 tcf onshore, 16 tcf offshore). Nearly all of the offshore proved reserves are located in the Central and Western Gulf of Mexico.

Table 2. U.S. Natural Gas Production: Federal and Non-Federal Areas FY2009-FY2013 (billion cubic feet)

Fiscal Year	U.S. Total	Non-Federal	Total Federal (% of U.S. Total)	Federal Offshore	Federal Onshore
2013	25,470	21,592	3,878 (15.2)	1,172	2,706
2012	25,208	20,938	4270 (16.9)	1,351	2,919
2011	23,539	18,953	4,586 (19.5)	1,668	2,918
2010	21,924	16,849	5,076 (23.2)	2,056	3,020
2009	21,612	16,241	5,372 (24.9)	2,205	3,167

Source: Federal data obtained from ONRR Statistics, http://www.onrr.gov (using sales year data), March 2014.

Notes: U.S. Fiscal Year Total data derived from EIA monthly production data in its publication "Natural Gas, U.S. Natural Gas Marketed Production," March 31, 2014, http://www.eia.gov.

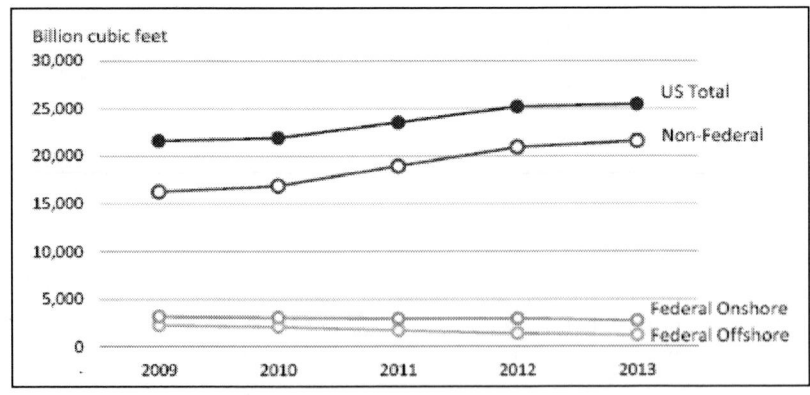

Source: Federal data obtained from ONRR Statistics, http://www.onrr.gov (using sales year data). Figure created by CRS.

Figure 2. U.S. Natural Gas Production: Federal and Non-Federal Areas FY2009-FY2013.

EIA Projections

While in the short-term, EIA estimates show oil production continuing to decline in federal offshore areas, EIA's longer-term estimates show a slight increase in federal offshore oil production overall, from 1.1 mbd in 2013 to 1.6-2.0 mbd in 2040.[7] Overall, the EIA projects U.S. oil production to rise from 7.4 mbd in 2013 to about 7.5 mbd by 2040 (essentially equal to 2013 production levels) after reaching 9.0 mbd in 2025.[8] According to these estimates, offshore production in 2040 could range from 21% to 27% of total U.S. crude oil production. (See *Table 3*.)

Offshore natural gas production is projected to reverse a years-long decline in 2015, with annual production rising as high as 2.9 tcf in 2040. Even though these projections are in calendar years, 2.9 tcf of natural gas is still likely more than a doubling of current offshore production (provided in fiscal years in the earlier sections of this report) but would only account for about a 7.7% share of total U.S. production in 2040. (See *Table 4*.)

Table 3. EIA Oil Production Projections
(million barrels per day)

Year	U.S. Offshore	U.S. Total
2025	n/a	9.0
2040	1.6-2.0	7.48

Source: EIA, Early Release Overview, 2014, Annual Energy Outlook, December 2013.

Table 4. EIA Natural Gas Production Projections
(trillion cubic feet per year)

Year	U.S. Offshore	U.S. Total
2025	n/a	31.93
2040	1.7-2.9	37.61

Source: EIA, Early Release Overview, 2014 Annual Energy Outlook, December 2013.

Oil and Natural Gas Lease Data for Federal Lands[9]

Currently, there are 113 million acres of onshore federal lands open and accessible for oil and gas development and about 166 million acres off-limits or inaccessible.[10] The Bureau of Land Management (BLM) is seeking to lease in areas where it anticipates fewer legal challenges; BLM also says it is addressing public concerns prior to a lease sale at a higher rate than in the past. In 2012, 56% of the onshore acreage under federal lease and 45% of federal onshore leases were not in production. Offshore, most of the 1.7 billion acres of federal water are no longer under leasing and development moratoria. The current five-year leasing program has lease sales scheduled in Western and Central Gulf of Mexico (GOM) and parts of Alaska.[11] In the offshore areas, 72% of the acreage is leased and 75% of the leases are not in production.

According to the BLM and the Bureau of Ocean Energy Management (BOEM), there are approximately 72.8 million acres of oil and gas leases in federal areas (onshore and offshore). About 37.0 million acres are located onshore and an additional 35.8 million acres are offshore. Approximately 11.1 million federal acres onshore and about 6.6 million federal acres offshore are producing commercial volumes. (See *Table 5.*)

Table 5. Oil and Gas Lease Data for Federal Lands, 2012

	Onshore	Offshore
Acreage under lease	37.0 million acres	35.8 million acres
Acreage with approved exploration or development plan (i.e., acreage in production or exploration)	16.3 million acres	10.1 million acres
Leased acres producing	11.1 million acres	6.6 million acres
Leased acres not in production or exploration	20.8 million acres	25.7 million acres
Number of Leases	49,213	6,621
Producing Leases (or with approved DOCD)[a]	27,300	1,611

Source: DOI, Oil and Gas Utilization—Onshore and Offshore, Report to the President, May 2012.

[a] A DOCD is a Development Operations Coordination Document that must be submitted for approval to BOEM before development activities begin.

Producing Acres

The number of federal producing acres may or may not be a function of how many acres are leased, and the number of acres leased may or may not correlate to production levels, but it is beyond the scope of this report to examine that issue thoroughly. In recent years, some members of Congress have proposed a $4/acre lease fee for non-producing leases. This proposal grew out of the efforts to open more public land and water (offshore) for oil and gas drilling and development when gasoline prices spiked in 2006-2008.

Some in Congress noted that there were many leases they believed were not being developed in a timely manner, while at the same time, others in Congress were advocating greater access to areas off-limits (such as ANWR and areas under leasing moratoria offshore). Higher rents for offshore leases were imposed by the Secretary of the Interior in 2009 to discourage holding unused leases and to move more leases into production, if possible. The escalation in rents is significant over time, as they rise from $7/acre to $28/acre (in year-8 forward) in water depths less than 200 meters, and increase from $11/acre to $44/acre (in year-8 forward) in water depths between 200 and 400 meters. However, there was no similar escalation for onshore leases, as they remain $1.50/acre for years 1-5, then rise to $2/acre thereafter.[12] A non-producing fee or an escalation of rents may not increase production but may reduce the ratio of producing leases to active leases. Thus, there might be fewer "idle" leases and acreage not in production or exploration. The BLM can re-lease acreage that has been relinquished or passed over at a future lease sale.

Applications for Permits to Drill (APDs)

Another major issue that Congress may address is streamlining the processing of applications for permits to drill (APDs). Some members contend that this would be one way to help boost energy production on federal lands. After a lease has been obtained, either competitively or noncompetitively, an application for a permit to drill must be approved for each oil and gas well. As noted in the Mineral Leasing Act, Section 226 (g), "no permit to drill on an oil and gas lease issued under this chapter may be granted without the analysis and approval by the Secretary concerned of a plan of operations covering proposed surface-disturbing activities within the lease area." The application form (APD form 3160-3) must include, among other things, a drilling plan, a surface use plan, and evidence of bond/surety coverage. The surface use plan should contain information on drillpad location, pad construction, the method for containment and waste disposal, and plans for surface reclamation.[13]

Prior to the Energy Policy Act of 2005 (P.L. 109-58, EPACT '05), a major concern that prompted the streamlining of permits debate was the lengthy timetable to process an APD. The BLM attributed the longer timelines to the rewriting of outdated Resource Management Plans (RMPs). There were several RMPs revised over the past decade. Leading up to the provisions in EPACT '05 that attempted to streamline the permitting process, the BLM announced, in April 2003, new strategies to expedite the APD process. The new strategies included processing and conducting environmental analyses on multiple permit applications with similar characteristics, implementing geographic area development planning for an oil or gas field or an area within a field, establishing a standard operating practice agreement that identifies surface and drilling practices by oil and gas operators, allowing for a block survey of cultural resources, promoting consistent procedures, and revising relevant BLM manuals.[14] EPACT '05 Section 366 (Deadline for Consideration of Application for Permits) provided a new timeline for BLM to process APDs.[15]

While the current Administration processed more APDs than it received from 2009-2011, it received far fewer applications over that period than the previous Administration had received from 2006-2008. Even though the number of pending applications has fallen steadily since 2008, the ratio of APDs pending to APDs processed was higher than during the period 2006-2008. In addition, there are 7,000 approved APDs that are not in the exploration or production stages (approved but not drilled).[16] The BLM expected to process more than 5,000 APDs in each of the fiscal years 2012 and 2013.

Table 6. Onshore Drilling Permits (FY2006-FY2011)

Fiscal Year	APDs Received	APDs Processed	APDs Pending
2011	4,278	5,200	4,309
2010	4,251	5,237	4,603
2009	5,257	5,306	5,589
2008	7,884	7,846	5,638
2007	8,370	8,964	5,600
2006	10,492	8,854	6,194

Source: U.S. Department of the Interior, Oil and Gas Utilization, Onshore and Offshore, May 2012.

It took an average of 307 days for all parties to process (approve or deny) an APD in 2011, but that has declined to an average of 194 days in 2013.[17] In 2006, it took the BLM an average of 127 days to process an APD, while in 2013 it took BLM 95 days. In 2006, the industry took an average of 91 days to complete an APD, but in 2013, the industry took 99 days. The BLM stated in its FY2012 and FY2013 budget justifications that overall processing times per APD rose to such high levels in 2011 because of the complexity of the process; now the permit process is improving, resulting in shorter timeframes.

Some critics of this lengthy timeframe highlight the relatively speedy process for permit processing on private lands. However, crude oil development on federal lands takes place in a wholly different regulatory framework than that of oil development on private lands.[18] State agencies permit drilling activity on private lands within their states, with some approving permits within 10 business days of submission. This faster approval rate does not necessarily diminish the additional work required by the state to address other state requirements. But often, some surface management issues are negotiated between the oil producer and the individual land/mineral owner. A private versus federal permitting regime does not lend itself to an "apples-to-apples" comparison.

Streamline Pilot

EPACT '05 also included a provision to initiate and fund (funding authorized through FY2015) a pilot program at seven BLM field offices in an effort to streamline the permitting process for oil and gas leases on federal lands. Results from the pilot project were published according to the timetable required by EPACT '05 (within three years after enactment). The conclusion was that the pilot made a difference in improving the processing times for APDs at the pilot offices overall and increased the number of environmental inspections. The BLM noted that the National Environmental Policy Act (NEPA) processing time for APDs and rights of way (ROW) applications fell from 81 to 61 days or roughly 25% due to "colocation" of agency staff. BLM reported that the number of environmental inspections went up by 78% from FY2006 to FY2007.[19] The BLM reported mixed results at the specific field offices. While some of the offices processed more permits in 2007 than they did in 2005, all the pilot sites reported more completed environmental inspections.[20]

Concerns over Non-Producing Leases

A number of concerns may arise in the oil and gas leasing process that could delay or prevent oil and gas development from taking place, or might account for the relatively large number of leases held in non-producing status. It should be noted that many leases expire without exploration or production ever occurring.

Below is a list of often-cited issues which, individually or in combination, are used to explain why more leases are not producing.

- Rig or equipment availability, particularly offshore;
- High capital costs and available capital;
- Skilled labor shortages;
- Leases in the development cycle (e.g., conducting environmental reviews, permitting, or exploring) but not producing;
- Legal challenges that might delay or prevent development;
- No commercial discovery on a lease tract;
- Holding leases (because of the lack of capital or as "speculators") to sell or "farm out" at a later date;
- Ability to secure extensions on non-producing leases;
- Securing and being able to hold large number of lease tracts, often contiguous, to maximize return on their investment; and
- The potential for inadequate coordination between the Department of the Interior's lease management and regulatory agencies (Bureau of Ocean Energy Management and Bureau of Safety and Environmental Enforcement) and other federal agencies to ensure protection of federal areas encompassing coastal and marine sanctuaries.

End Notes

[1] For a broader analysis of offshore oil and gas leasing and resources, see CRS Report R40645, U.S. Offshore Oil and Gas Resources: Prospects and Processes, by Marc Humphries and Robert Pirog.

[2] For more information on U.S. oil development, see CRS Report R43148, An Overview of Unconventional Oil and Natural Gas: Resources and Federal Actions, by Michael Ratner and Mary Tiemann; CRS Report R41132, Outer Continental Shelf Moratoria on Oil and Gas Development, by Curry L. Hagerty; and CRS Report R43429, Federal Lands and Natural Resources: Overview and Selected Issues for the 113th Congress, coordinated by Katie Hoover.

[3] The early data (1980 and 1990s) were taken from annual Mineral Revenue reports. The data used at that time were accounting data which are considered by the Office of Natural Resources Revenue as not very reliable. The more useful production volume data provided by ONRR now are based on fiscal year sales data.

[4] EIA, U.S. Crude Oil and Natural Gas Proved Reserves, 2011, August 2013, http://www.eia.gov.

[5] U.S. natural gas production on federal lands fell from about 7 trillion cubic feet in FY2003 to about 4 trillion cubic feet in FY2013.

[6] EIA, U.S. Crude Oil and Natural Gas Proved Reserves, 2011, August 2013, http://www.eia.gov. Dry gas is marketed production less extraction losses.

[7] EIA, Early Release Overview, 2014, Annual Energy Outlook, December 2013.

[8] Ibid.

[9] 2013 data from BLM was not available at the time of this writing.

[10] U.S. Depts. of the Interior, Agriculture, and Energy, Inventory of Onshore Federal Oil and Natural Gas Resources and Restrictions to Their Development (Phase III), May 2008, available on the BLM website at http://www.blm.gov/ wo/st/en/prog/energy/oil_and_gas/ EPCA_III.html. The availability of public lands for oil and gas leasing can be divided into three categories: lands open under standard lease terms, open to leasing with restrictions, and closed to leasing. Areas are closed to leasing pursuant to land withdrawals or other mechanisms. Much of this withdrawn land consists of wilderness areas, national parks and monuments, and other unique and environmentally sensitive areas that are unlikely to ever be reopened to oil and gas leasing. Some lands are closed to leasing pending land use planning or NEPA compliance, while other areas are closed because of federal land management decisions on endangered species habitat or historical sites. Some of those restricted areas may be opened by future administrative decisions.

[11] The Eastern GOM is under a leasing moratoria until 2022 under the Gulf of Mexico Energy Security Act, and the North Aleutian Basin of Alaska was withdrawn from leasing under an executive order by the current Administration.

[12] DOI, Oil and Gas Lease Utilization, Onshore and Offshore, Updated Report to the President, May 2012, p.18.

[13] U.S. Department of the Interior, Bureau of Land Management (BLM), Surface Operating Standards and Guidelines for Oil and Gas Exploration and Development, The Gold Book, Fourth Edition-Revised 2007, p. 8.

[14] DOI/BLM Instruction Memorandum No. 2003-152, Application for Permit to Drill Process Improvement#1- Comprehensive Strategies, April 14, 2003.

[15] Within 10 days of receiving the application from the operator, BLM shall notify the operator as to whether the application is complete and also schedule a site visit. If the application is not complete, the operator then has 45 days to submit additional information to BLM to complete the application or the application is returned to the operator. Within 30 days of receiving a completed application the BLM will approve or defer the application. If deferred, the operator has up to two years to take specified actions to complete the application or face the possibility of being denied a permit.

[16] U.S Department of the Interior, Oil and Gas Lease Utilization, Onshore and Offshore, Updated Report to the President, May 2012, p. 14.

[17] Bureau of Land Management, "Average Application for Permit to Drill (APD) Approval Timeframes: FY2005- FY2012," http://www.blm.gov/wo/st/en/prog/energy/oil_and_gas/ statistics/apd_chart.html.

[18] Under the Federal Land Policy and Management Act (FLPMA), Resource Management Plans or Land Use Plans (43 U.S.C. 1712) are required for tracts or areas of public lands prior to development. The Bureau of Land Management (BLM) must consider environmental impacts during land-use planning when RMPs are developed and implemented. RMPs can cover large areas, often hundreds of thousands of acres across multiple counties. Through the land-use planning process, the BLM determines which lands with oil and gas potential will be made available for leasing.

[19] Bureau of Land Management, BLM Year Two Report, Section 365 of EPACT 2005 Pilot Project to Improve Federal Permit Coordination, February 2008.

[20] Ibid.

In: Oil and Gas Development ...
Editor: Aaron Spearing

ISBN: 978-1-63321-779-9
© 2014 Nova Science Publishers, Inc.

Chapter 3

OIL AND GAS LEASE UTILIZATION, ONSHORE AND OFFSHORE, UPDATED REPORT TO THE PRESIDENT[*]

U.S. Department of the Interior

INTRODUCTION AND EXECUTIVE SUMMARY

On March 11, 2011, President Obama directed the Department of the Interior (Department) to determine the acreage of public lands that had been leased to oil and gas companies but remain undeveloped, noting that companies should be encouraged to produce energy from leases that they are holding. The Report reached several important conclusions: First, the Department has offered substantial acreage for leasing and resource development, but much of this acreage has not been leased by industry. Second, tens of millions of acres that are currently under lease remain idle. Because these areas are not undergoing exploration, development, or production, taxpayers are not getting the full advantage of America's resource potential.

Soon after the release of the Report, the President released a *Blueprint for a Secure Energy Future,* a comprehensive energy strategy to secure America's energy future by producing more conventional energy at home while working to reduce our dependence on oil by leveraging cleaner,

---

[*] This is an edited, reformatted and augmented version of report issued May 2012.

alternative fuels and greater efficiency. Included in the *Blueprint* were a number of steps to encourage safe and responsible domestic energy production both onshore and offshore – including steps to encourage diligent development.

This Report updates the Department's 2011 Report, outlines some of the policies that are being implemented, consistent with last year's findings and the *Blueprint*, and describes some of the geographic areas that best reflect the promise of this Nation's onshore and offshore energy resources.

For both onshore and offshore, the report describes:

Leasing Trends and Unused Acreage

In 2011 alone, the Department offered about 21 million offshore acres for oil and gas development. And, we are continuing to offer substantial acreage where the oil is. Next month, on June 20, 2012, the Department will offer more than 38 million acres as part of a lease sale in the Central Gulf of Mexico, an area estimated to hold close to 31 billion barrels of oil and 134 trillion cubic feet of natural gas that are currently undiscovered and technically recoverable – and home to some of the most promising deepwater prospects. Onshore, the Department offered 1.2 million onshore acres for lease during fiscal year 2011, as well as an additional 3 million acres in the first quarter of FY 2012 alone.

- Offshore, Gulf of Mexico lease sales typically offer virtually all unleased available acreage.
- Onshore, industry expresses interest in leasing lands. Thus, industry plays a large role in driving the amount of parcels/acres being offered and leased.

There are approximately 26 million leased acres offshore and over 20 million leased acres onshore that are currently idle – that is, not undergoing exploration, development, or production.

Offshore: As of May 2012, nearly 72 percent of the area on the Outer Continental Shelf (OCS) that companies have leased for oil and gas development – totaling 26 million acres – are not producing or not subject to pending or approved exploration or development plans.

In the Gulf of Mexico, which holds the largest volume of undiscovered technically recoverable resource (UTRR) on the OCS, 32 million acres are under lease. However, only approximately 10 million acres have approved exploration or development plans, and only 6.4 million of these acres are in production. Leased areas in the Gulf of Mexico – that are not producing or not subject to pending or approved exploration and development plans – are estimated to contain 17.9 billion barrels of UTRR oil and 49.7 trillion cubic feet of UTRR natural gas.

Onshore: As of December 31, 2011, approximately 56 percent of total acres of public land under lease in the Lower 48 States – totaling approximately 20.7 million acres - are not undergoing either production nor exploration activities.

As of September 30, 2011, there are over 7,000 approved permits to drill on public and Indian lands that have not yet been acted on by companies.

Roughly 76 percent of the onshore acres offered for sale between October 1, 2010, and September 30, 2011, were bid on and sold for oil and gas activities.

Policy Actions to-Date

The Department is committed to providing companies with incentives for rapid development of oil and gas production from existing and future leases. For offshore, reforms since 2009 have included establishment of escalating rentals, restructured initial lease terms in certain water depths, and higher minimum bids for lease sales, all to encourage diligent development and fair return to taxpayers. Onshore, reforms have focused on new leasing policies that establish a more orderly, open, and environmentally sound process for efficient development of oil and gas resources on public lands. The Department also is developing an Advanced Notice of Proposed Rulemaking (ANPR) seeking input on potential incentives to encourage timely development of unused onshore leases.

Moving Forward – Spotlight on Key Prospects

This Report also describes some of the most promising areas for development, both onshore and offshore. Because of the significant resource potential that these areas demonstrate, they are central to the focus of DOI and its bureaus as we advance our efforts to pursue safe and responsible exploration and development both offshore and onshore.

This report was prepared by the Department's Office of Policy Analysis; data sets used in the report were compiled by the United States Geological Survey (USGS), Bureau of Land Management (BLM), and Bureau of Ocean Energy Management (BOEM).

OFFSHORE OIL AND GAS LEASES

Background

The BOEM manages energy and mineral resources on the Nation's OCS to help meet the energy demands and other needs of the Nation while balancing development with the protection of the human, marine, and coastal environments. For the purposes of this report, the areas of the OCS leased for offshore oil and gas development can be broadly categorized as:

- Total acres under lease;
- "Active leases," or leased areas that are subject to exploration, development, or production; and
- "Inactive leases" or leased areas that are not producing nor currently covered by an approved exploration or development plan. These areas may be subject to certain activities such as geophysical and geotechnical analysis, including seismic and other types of surveys.

The regulatory process governing offshore oil and gas exploration and development can be briefly summarized as follows:

- Leasing: The Secretary of the Interior (Secretary) must prepare a Five-Year OCS Oil and Gas Leasing Program consisting of a schedule of oil and gas lease sales indicating the size, timing, and location of proposed leasing activity the Secretary determines will best meet national energy needs. Preparing a Five-Year program involves extensive public comment and requires the Secretary to balance many factors including the potential for the discovery of oil and natural gas, the potential for environmental damage, and the potential for adverse effects on the coastal zone. An area must be included in the Five-Year program in order to be offered for leasing. There is an additional public process for each lease sale to determine whether to hold the lease sale or modify the sale area, and what terms and conditions to apply to leases. The BOEM is currently finalizing the OCS Oil and Gas Leasing Program for 2012-2017, described in more detail below. The Department released the Proposed Five-Year Program in November 2011.

- Exploration: Exploration activities include geophysical exploration and exploratory drilling. Prior to conducting any exploratory drilling activity on a lease, an Exploration Plan (EP) must be submitted to BOEM for approval. The EP describes exploration activities planned by the operator for a specific lease(s), the timing of these activities, information concerning drilling rigs, the location of each well, and other relevant information.
- Development: Before an operator can begin development or production activities, a Development and Production Plan (DPP), or a Development Operations Coordination Document (DOCD) must be submitted for approval to BOEM.[1] The plan describes a schedule of development activities, platforms, or other facilities including environmental monitoring features and other relevant information. As with EPs, BOEM can require modification of a plan.

The BOEM is committed to conducting efficient and thorough reviews of these plans, and to ensuring that the process is rigorous, efficient, and transparent. BOEM works collaboratively with industry throughout the review of plans, with the goals of ensuring that operators comply with BOEM's heightened operational and environmental standards and that the review process is efficient. The BOEM currently conducts a full, site-specific environmental assessment on all deepwater exploration and development plans. The Bureau of Safety and Environmental Enforcement (BSEE) reviews and approved APD's that leads to production.

Leasing Trends and Unused Acreage

The Administration is committed to making the offshore areas with the most substantial resources available to companies, and to incentivizing diligent development of leases. The *Deepwater Horizon* blowout and oil spill made clear the tremendous human and environmental costs that can come from deepwater oil and gas drilling when proper safeguards are not followed. In light of lessons learned, BSEE has taken significant actions to reform and strengthen our offshore drilling safety regime to increase safety and preparedness. Consistent with heightened standards, offshore oil and gas exploration and development is moving forward, and industry is making significant investments in the Gulf of Mexico and elsewhere.

On December 14, 2011, BOEM held Western Gulf of Mexico Lease Sale 218, the first sale since *Deepwater Horizon,* which attracted over 240 bids on 191 tracts, with nearly $338 million in total high bonus bids -- about $100 million more than the average for Western Gulf sales over the previous decade. The Administration has announced that BOEM will hold Consolidated Central Gulf of Mexico Lease Sale 216/222 on June 20, 2012. Sale 216/222 will make available all unleased, eligible areas in the Central Gulf of Mexico, a planning area that BOEM estimates contains close to 31 billion barrels of oil and 134 trillion cubic feet of natural gas that are currently undiscovered and technically recoverable. The Central Gulf alone is estimated to hold about a third of the undiscovered resources of the OCS.

Sale 216/222 is the last remaining sale scheduled in the 2007 – 2012 OCS Oil and Natural Gas Leasing Program. As the President discussed in his State of the Union, the Department is finalizing the next Five-Year Program for 2012-2017, which, as proposed, would make more than 75 percent of estimated undiscovered technically recoverable oil and gas resources on the OCS available for development. The Proposed Outer Continental Shelf Oil and Gas Leasing Program for 2012-2017 schedules 12 potential lease sales in the Gulf of Mexico, as well as three potential sales in Alaska's Cook Inlet, Chukchi Sea, and Beaufort Sea.

Although BOEM continues to make significant new resources available, our analysis shows that significant existing, leased resources remain unexplored and undeveloped. Overall, as of May 2012, nearly 72 percent (totaling 25.7 million acres) of acres on the Outer Continental Shelf that companies have leased for oil and gas development are not producing or not subject to pending or approved exploration and development plans. This compares to over 72 percent as of March 1, 2011. In all Federal areas of the Gulf of Mexico, approximately 69 percent of acreage that is currently under lease is not producing or subject to pending or approved exploration or development plans. In the Gulf of Mexico, which holds the largest volume of undiscovered technically recoverable resource (UTRR) on the OCS, approximately 31.9 million acres are under lease. However, only approximately, 9.8 million acres have approved exploration or development plans, and 6.4 million of the 9.8 million acres are in production. Leased areas in the Gulf of Mexico that are not producing or not subject to pending or approved exploration and development plans are estimated to contain 17.9

billion barrels of UTRR oil and 49.7 trillion cubic feet of UTRR natural gas *(see Table 2)*. In addition, offshore permitting is nearly back to pre-*Deepwater Horizon* levels, while at the same time, we have instituted the largest offshore drilling reforms in U.S. history to make sure that development happens safely and responsibly. For example, 67 drilling permits for wells for new deepwater wells (more than 500 feet deep) in the 12 months ending April 19, 2012 – just three fewer than in the same period (from April 2009-2010) before the *Deepwater Horizon* explosion. Overall, since these new standards were put into place, the Administration has approved over 600 permits for activities at hundreds of wells in the Gulf of Mexico alone.

Table 1. U.S. Offshore Lease Activity (As of May 2012)

Region	Total Leased Acres	Inactive Lease Acres	Active Lease Acres
Gulf of Mexico	31,864,710 [5,902 leases]	22,033,940 [3,918 leases]	9,830,770[3] [1,984 leases]
Pacific[1]	241,023 [49 leases]	23,354 [6 leases]	217,669 [43 leases]
Alaska2	3,723,465 [670 leases]	3,650,974 [656 leases]	72,491 [14 leases]
Total Offshore	35,829,198 [6,621leases]	25,708,268 [4,580 leases]	10,120,930 [2,041 leases]

[1] No lease sales have been held in the Pacific region since 1984.
[2] Approximately three-quarters of leased Alaska acreage is subject to litigation (challenging 2008 Chukchi Sea Lease Sale 193).
[3] The BOEM held the Western Gulf of Mexico Oil and Gas Lease Sale 218 on December 14, 2011, and subsequently completed its required evaluation. The BOEM awarded 181 leases on tracts covering 1,036,205 acres to the successful high bidders who participated in the sale. These leases were awarded in 2012; thus, there has not yet been sufficient time for those leases to enter into the "active lease" category—as defined in this report.

In the Central and Western Gulf of Mexico, approximately 53 million acres were offered for lease in 2009, of which 2.7 million acres were bid on and sold. In the Central Gulf, approximately 37 million acres were offered in 2010, of which 2.4 million acres were bid on and sold; in the 2011 Western Gulf lease sale, over 21 million acres were offered, with over a million acres sold *(see Table 3)*.

Table 2. U.S. Offshore Active Leases (As of May 2012)

Region	Total Active Acres	Acres with Approved Development Plans	Acres with Approved Exploration Plans	Producing Acres[3]
Gulf of Mexico	9,830,770 [1,984 leases]	7,479,805 [1,563 leases]	2,350,965 [421 leases]	6,409,169
Pacific[1]	217,669 [43 leases]	217,669 [43 leases]	0	217,669
Alaska[2]	72,491 [14 leases]	21,254 [5 leases]	51,237 [9 leases]	10,414
Total	10,120,930 [2,041 leases]	7,718,728 [1,611 leases]	2,402,202 [430 leases]	6,637,252

[1] No lease sales have been held in the Pacific region since 1984.
[2] Approximately three-quarters of leased Alaska acreage is subject to litigation (challenging 2008 Chukchi Sea Lease Sale 193).
[3] Producing acreage is a subset of the acreage subject to approved development plans/DOCDs.

Table 3. Gulf of Mexico Acreage Offered and Leased

	2009		2010		2011	
Region	Acres Offered	Acres Leased	Acres Offered	Acres Leased	Acres Offered	Acres Leased
Central Gulf	34,594,940	1,784,242	36,957,957	2,369,101	0	0
Western Gulf	18,393,357	884,167	0	0	21,010,305	1,036,205[2]
Total Gulf[1]	52,988,297	2,668,409	36,957,957	2,369,101	21,010,305	1,036,205

[1] Gulf of Mexico lease sales typically offer virtually all unleased available acreage. With the exception of a small number of leases in special circumstances (e.g., within the boundaries of the Flower Garden Banks National Marine Sanctuary), every unleased acre gets reoffered every year. This includes leases which return to the Federal inventory due to: expiration of the lease term; relinquishment; end of production; and acres that were not leased in a previous sale.
[2] All leases of Sale 218 became effective in 2012.

Policy Actions to-date: Encouraging Diligent Development

Consistent with the Administration's *Blueprint for a Secure Energy Future*, BOEM has implemented administrative reforms to ensure fair return to taxpayers and encourage diligent development. These measures include:

- **Increasing rental rates to encourage faster exploration and development of leases:** In the Gulf of Mexico, during the initial term of a lease and before the commencement of royalty-bearing production, the lessee pays annual rentals which either step-up by almost half after year 5 – for leases in water 400 meters or deeper – or escalate each year after year 5 – for leases in less than 400 meters of water. The primary use of step-up and escalating rentals is to encourage faster exploration and development of leases, or earlier relinquishment when exploration is unlikely to be undertaken by the current lessee. Rental payments also serve to discourage lessees from purchasing marginally valued tracts, and provide an incentive for the lessee to drill the lease or to relinquish it, thereby giving other market participants an opportunity to acquire these blocks. In March 2009, in addition to implementing escalating rental rates, BOEM raised the base rental rates for years 1-5 *(see Appendix 1)*.
- **Tiered durational terms to incentivize prompt exploration and development:** Gulf of Mexico leases in certain water depths (400-1600 meters) are now structured to provide for relatively short initial periods, but followed by an additional period under the same lease if the operator drills a well during the initial period. The initial periods are graduated by water depth to account for technical differences in operating at various water depths *(see Appendix 1)*. In addition, the Bureau of Safety and Environmental Enforcement (BSEE) recently informed lessees of a decision from the Department's Office of Hearings and Appeals that reaffirms the requirement that lessees demonstrate a commitment to produce oil or gas in order to be eligible for lease expiration suspensions.
- **Increased minimum bid:** In 2011, BOEM increased the minimum bid for tracts in at least 400 meters of water in the Gulf of Mexico to $100 per acre, up from $37.50, to ensure that taxpayers receive fair market value for offshore resources and to provide leaseholders with additional impetus to invest in leases that they are more likely to develop. Analysis of the last 15 years of lease sales in the Gulf of

Mexico showed that deepwater leases that received high bids of less than $100 per acre, adjusted for energy prices at time of each sale, experienced virtually no exploration and development drilling.

Strong industry response to Western Gulf of Mexico Lease Sale 218, held in December 2011, suggests that these incentives encouraged significant, targeted investment in areas most likely to lead to production. The sale attracted over 240 bids on 191 tracts, with nearly $338 million in total high bonus bids – about $100 million more than the average for Western Gulf sales over the previous decade.

In addition to the incentives described above, the Department is working to reduce barriers to development elsewhere in the Gulf of Mexico. On February 20, 2012, the Departments of the Interior and State joined officials from the Government of Mexico to sign an agreement on the exploration and development of transboundary oil and natural gas reservoirs along the United States–Mexico maritime boundary in the Gulf of Mexico.

This Agreement, when implemented, will make an additional 1.5 million acres of the U.S. Outer Continental Shelf will be made more accessible for exploration and production activities.

Finally, it should be noted that the Department has been moving forward with the processing of plans to explore leases in the Beaufort and Chukchi Seas in Alaska.

Pursuant to an Executive Order that the President issued in July 2011, an Interagency Working Group on Coordination of Domestic Energy Development and Permitting in Alaska has been set up. This working group has been coordinating Federal agency reviews of industry requests to engage in exploratory drilling activity in the Arctic.

The result has been, for the first time in history, a government-wide approach to ensuring that safety and environmental issues are being addressed in both a comprehensive and timely fashion.

OFFSHORE SPOTLIGHT: CENTRAL GULF OF MEXICO

The Gulf of Mexico has the largest untapped resource potential in the United States' Outer Continental Shelf (OCS)and currently supplies over a quarter of the Nation's domestic oil production. Within the Gulf of Mexico, the Central Gulf has by far the greatest potential, particularly in the deepwater areas, where there are already several world class producing basins and a number of significant new discoveries have been made.

According to BOEM's *Assessment of Undiscovered Technically Recoverable Oil and Gas Resources of the Nation's Outer Continental Shelf*, issued in 2011, it is estimated that the Gulf of Mexico holds about 55 percent of the potential resources located on the OCS, in barrels of oil equivalent. The Central Gulf alone is estimated to have over 30 billion barrels of oil and 133.9 trillion cubic feet of natural gas yet to be discovered, nearly double the resource potential of any other OCS planning area *(see Figure 5)*.

High resolution, depth migrated 3-D seismic data are revealing new realms for exploration, including untapped subsalt plays as found in the Lower Tertiary trend, also called the Wilcox play, which has emerged as one of the world's leading exploration plays due to significant, recent discoveries with a high concentration in the Central Gulf.

Figure 6 illustrates the geographic distribution of 15 Lower Tertiary discoveries, including the 3 announced discoveries found in 2009 and 12 significant discoveries that have taken place in BOEM-designated fields. The map also shows the approximate spatial extent of this trend, which may cover over 30,000 square miles (77,700 km2) at an average depth of 27,500 ft. (8,382 m) subsea.

New and rapidly advancing drilling technology and techniques, including innovations in directional drilling, make it possible to explore new frontiers by enabling drilling crews to reach around salt structures to access new reservoirs.

The Administration's comprehensive energy strategy reflects the importance of encouraging exploration and development in the Central Gulf of Mexico. As the President noted, BOEM will hold the consolidated Central Gulf of Mexico Lease Sale 216/222 in New Orleans on June 20, 2012, offering all available unleased acreage.

The Proposed OCS Oil and Gas Leasing Program for 2012-2017 schedules 12 lease sales in the Gulf of Mexico, including annual, area-wide lease sales in the Central Gulf and Western Gulf. The Proposed Program schedules the majority of sales in these areas, where resources are best known and where the infrastructure that exists in the Gulf to support the offshore oil and gas industry, including infrastructure to bring resources to market and respond to emergencies, is the most mature and well developed in the country.

ONSHORE OIL AND GAS LEASES

Background

The BLM is the leasing agent for all energy minerals on approximately 700 million acres of Federal lands, and supervises operational activities on leases on Indian lands held in trust by the United States.[2] The lands that are leased by BLM for oil and gas development can be broadly categorized as:

- *Active leases*: Areas with ongoing exploration or production activities. Exploration activities include exploratory and geophysical exploration.
- *Inactive leases*: Areas with no ongoing exploration or production activities.

The regulatory process for onshore oil and gas exploration and development includes:

- Leasing: Unlike the offshore process, BLM accepts "expressions of interest" from industry for lands to be placed on future oil and gas lease sales. "Expression of interest" areas are then evaluated by BLM to determine their leasing eligibility and conditions. Lease sales are then held that allow potential developers to bid on areas they are interested in for exploration and development.
- Exploration and Development: Operators are required to submit an "Application for Permit to Drill" (APD) that includes details such as the well plat, drilling plan, evidence of bond coverage, and operator certification. The APD must also include a Surface Use Plan of Operations. The Surface Use Plan of Operations must describe the proposed project in a narrative, as well as on maps and diagrams.

Leasing Trends and Unused Acreage

As of March 31, 2012, approximately 56 percent (20.8 million acres) of total onshore acres under lease on public lands in the Lower 48 States were conducting neither production nor exploration activities[3] *(see Table 4).*

- This represents a slight change from the 2011 Report to the President, when approximately 57percent of total onshore acres under lease on public lands in the Lower 48 States were conducting neither production nor exploration activities.

Since the mid-1990s, about 30 percent of leased acres have been in producing status *(see Figure 2)*.

Roughly 76 percent of the onshore acres offered for sale between October 1, 2010, and September 30, 2011, were bid on and sold for oil and gas activities (see Table 5).

- This compares to 23 percent that were bid on and sold between October 1, 2009, and September 30, 2010.
- Acreage not sold during a lease sale remains available "over-the-counter" for up to two years.

In calendar year 2011, BLM held 32 lease sales covering 4.4 million acres, including three of the top five largest sales in the agency's history (in Montana, Utah and Wyoming).

- This compares to calendar year 2010 when the BLM held 29 oil and gas lease sales covering 3.2 million acres.
- The BLM expects to hold 30 more lease sales in calendar year 2012.

The number of APDs awaiting approval has been reduced by 24 percent over the last three years, from 5,638 pending approval at the end of FY 2008 to 4,309 pending approval at the end of FY 2011. (see Figure 1 and Table 6).

- The BLM approved 4,725 APDs during FY 2011, and expects to process 5,500 APDs in FY 2012, and another 5,500 in FY 2013.

Over the last year, the number of "approved-but-not-drilled" APDs (on Federal and Native American lands) remained approximately 7,000.

Table 4. Onshore Oil and Gas Lease Activity - Lower 48 States (As of March 31, 2012)

Lease Category	Acres Under Lease	Percent of Total Acres	Number of Leases	Percent of Total Leases
Production & Exploration	16,280,081	44	27,307	55
Not in Production or Exploration	20,761,372	56	21,906	45
Total	37,041,453	100	49,213	100

Source: BLM LR2000 data; BLM AFMSS data.

Note: These data are subject to State verification and may differ from those shown in the BLM's Public Land Statistics publication. Further, due to an artifact of an ongoing data-system migration, the figures likely understate the number of leases with approved geophysical exploration permits; the BLM is conducting additional review, and anticipates that the figures may be revised in a future publication.

Table 5. Onshore Oil and Gas Lease Sales – Fiscal Year Statistics

	Acres Offered	Acres Sold	Percent Acres Sold	Parcels Offered	Parcels Sold	Percent Parcels Sold
FY 2009	3,803,635	1,819,234	48	3,127	1,874	60
FY 2010	3,239,086	739,954	23	1,636	1,003	61
FY 2011[1]	1,158,808	880,895	76	1,440	1,253	87

[1] In the first quarter of FY2012 the Department has already offered an additional 3 million onshore acres in the National Petroleum Reserve – Alaska.

Table 6. Onshore Drilling Permits (FY 2001-FY 2011)

Fiscal Year	APDs Received	APDs Processed	APDs Pending at End of Year
2001	4,819	4,266	5,638
2002	4,585	5,830	4,393
2003	5,063	5,143	4,313
2004	6,979	7,351	3,941
2005	8,351	7,736	4,556

Fiscal Year	APDs Received	APDs Processed	APDs Pending at End of Year
2006	10,492	8,854	6,194
2007	8,370	8,964	5,600
2008	7,884	7,846	5,638
2009	5,257	5,306	5,589
2010	4,251	5,237	4,603
2011	4,278	5,200	4,309

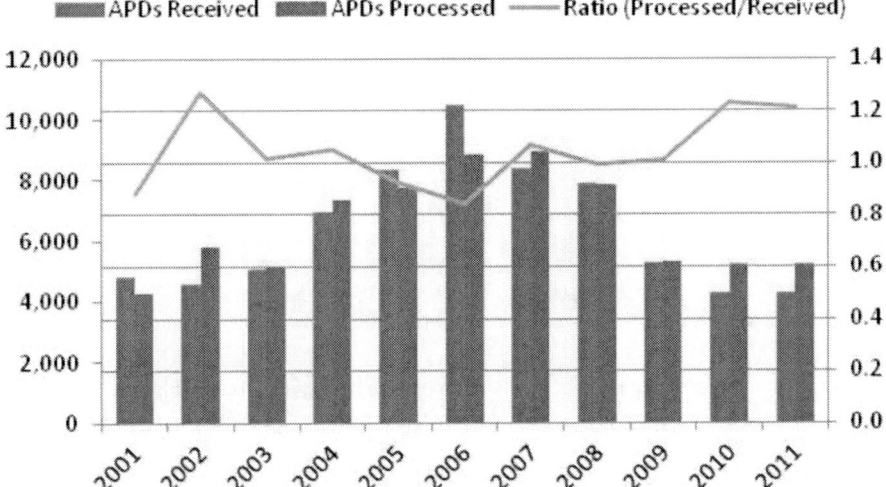

Note: Given potential backlogs, the number of APDs processed might exceed the number of APDs received for any given year.

Figure 1. Onshore APDs Received and Processed (FY 2001-FY 2011).

Policy Actions to-Date: Focusing on an Open and Efficient Process

The BLM is implementing oil and gas leasing reforms to ensure that oil and gas lease sales will offer parcels in appropriate locations and avoid the contention and litigation that have characterized many development proposals over the past several years. The BLM staff now work with local communities and address conflicts prior to lease sales, so that leasing activities—and the jobs that they generate—can move forward without being held up by protests or litigation. Master Leasing Plans are also being implemented to incorporate environmental concerns and help guide industry to lower-conflict areas for development.

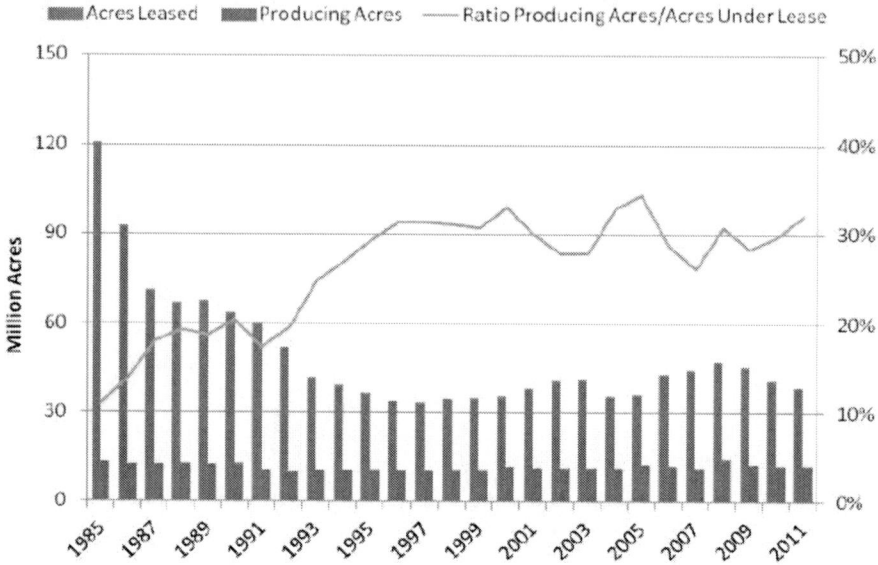

Figure 2. Onshore Producing Acres as a Percentage of Leased Acres, FY 2001 – FY 2011.

Efforts to better plan for our lease sales are producing successful results. In 2009, nearly 48 percent of all new proposed oil and gas parcels were being protested. That figure declined to 41 percent during 2010. Since the implementation of leasing reforms in early 2011, the number has declined further to 36 percent for fiscal year 2011. These BLM leasing reforms are providing more opportunity to move forward with appropriate oil and gas production activities on public lands while ensuring the protection of important natural resources. Further, BLM has addressed protests prior to lease sales at a higher rate providing potential bidders with a more secure, predictable pathway to development.

The Department also is developing an Advanced Notice of Proposed Rulemaking (ANPR) seeking input on potential incentives to encourage timely development of unused onshore leases. The ANPR will present potential royalty rate options and solicit comments regarding those options. One of the options in the ANPR will concern the extent to which incentives could be used to encourage production earlier in the life of the lease. A future rulemaking would apply to new onshore leases only.

ONSHORE SPOTLIGHT: BAKKEN SHALE OIL

The BLM Montana/Dakotas State Office manages nearly 2 million subsurface acres of mineral estate in North Dakota, where the Bakken play is currently experiencing a significant boom in oil production, making North Dakota the Nation's 4th largest oil producing State. The Bakken play in North Dakota and Montana is projected by industry experts to last through the next four decades and could surpass both Alaska and California in oil production within the next ten years.

Although the majority of the Bakken play is under private lands, BLM serves as the leasing authority for all Federal fluid minerals in the Bakken play, stipulating all leases with appropriate measures to maintain environmental quality and human safety. The BLM holds quarterly competitive lease sales, returning half of the revenue to the State. In July 2011, BLM held the second highest-grossing lease sale for Federal minerals, which brought a total of $66 million in revenue. The January 2012 lease sale brought nearly $36 million.

The BLM is the lead agency for permitting, inspection and enforcement activities for Federal and tribal mineral resources in the Bakken. Using BLM technical, geologic, and engineering expertise, along with input from the surface managers (BLM, USFS, BIA, Tribes), BLM is responsible for analyzing and responding to all Federal and Indian drilling applications. Drilling applications increased 500 percent over the past five years; half of this increase has occurred on Indian minerals. Since 2007, applications to drill on the Ft. Berthold Reservation have increased from 0 to 175. More than $3 million in drilling permit fees were collected in FY 2011.

Along with BLM's drilling oversight and surface protection requirements, BLM provides inspection and enforcement for production accountability. These workloads have increased 450 percent in the past five years, with some of these wells producing up to 1000 barrels of oil per day. Federal royalties in North Dakota totaled over $318 million in FY 2011. Indian wells generated $180 million revenues in FY 2011. Since 2007, on Fort Berthold alone, revenues have gone from less than $1 million to $134 million per year.

APPENDIX 1. BOEM – DEVELOPMENT INCENTIVES INCLUDED IN THE MARCH 2010 CENTRAL GULF OF MEXICO LEASE SALE

Lease Terms

The initial period of the lease is the principal diligence tool for OCS leases. A lease expires at the end of the initial period unless it is producing, or approved drilling operations are being conducted or a suspension of production or operations has been approved by BSEE.

In March 2010, Secretary Salazar shortened the initial period for future leases in certain water depths to the length shown in Table 8. Leases in 400 to 1,600 meters of water can obtain a three-year extension if the operator has spudded a well and submitted the information for BOEM District Manager confirmation.[4] The BOEM lease analysis indicated that, for leases in 800 to 1,600 meters of water, a seven-year lease term would be sufficient for an operator to evaluate seismic data and to commence drilling.

Table 8. Offshore Lease Terms

Water Depth (in meters)	Primary Lease Term	Extensions for Wells Spudded
< 400	5 years	3 years*
400 to < 800	5 years	3 years
800 to < 1600	7 years	3 years
1600+	10 years	n/a

* In less than 400 meters of water, a qualifying well must target hydrocarbons below 25,000 feet Total Vertical Depth Subsea.

Rental Rates

Rental rates are per-acre payments for leases that are paid annually until the start of royalty-bearing production. Prior to March 2009, BOEM's standard rental rates were $6.25 per acre in water less than 200 meters deep, and $9.50 per acre in water depths of 200 meters or more with escalation after year 5 for leases in water less than 400 meters in depth. In March 2009, Secretary Salazar increased the base rental rates for new leases offered in Gulf of

Mexico lease sales, and established step-up rentals on leases in water depths of 400 meters or more as an incentive for early exploration. These new rental rates are shown in Table 9, below.

**Table 9. Off and Onshore Lease Rental Rates
(per acre or fraction of an acre)**

	Year 1 through Year 5	Year 6	Year 7	Year 8 and Onward
Onshore	$1.50	$2	$2	$2
Offshore (by water depth)				
< 200 m	$7	$14	$21	$28
200 m to < 400 m	$11	$22	$33	$44
400+ m	$11	$16	$16	$16

APPENDIX 2. MAPS

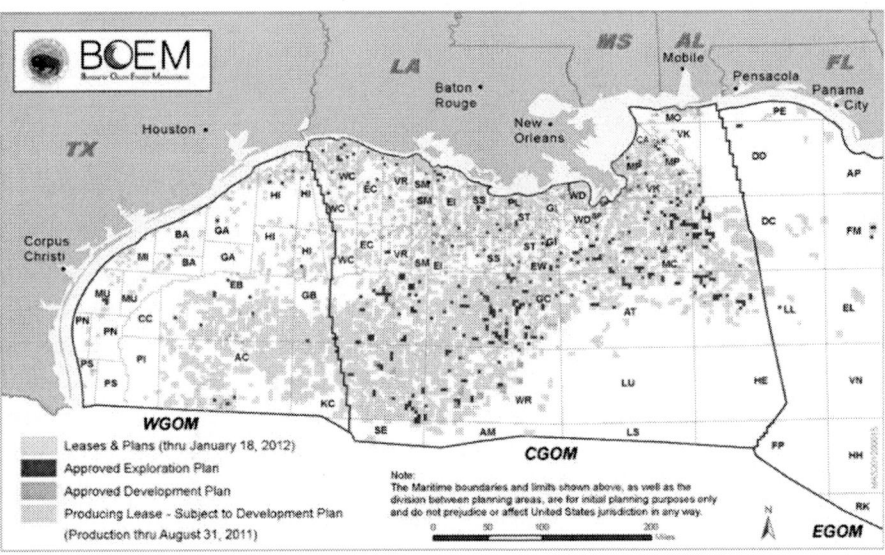

Figure 3. Gulf of Mexico OCS Leases (as of January 2012).

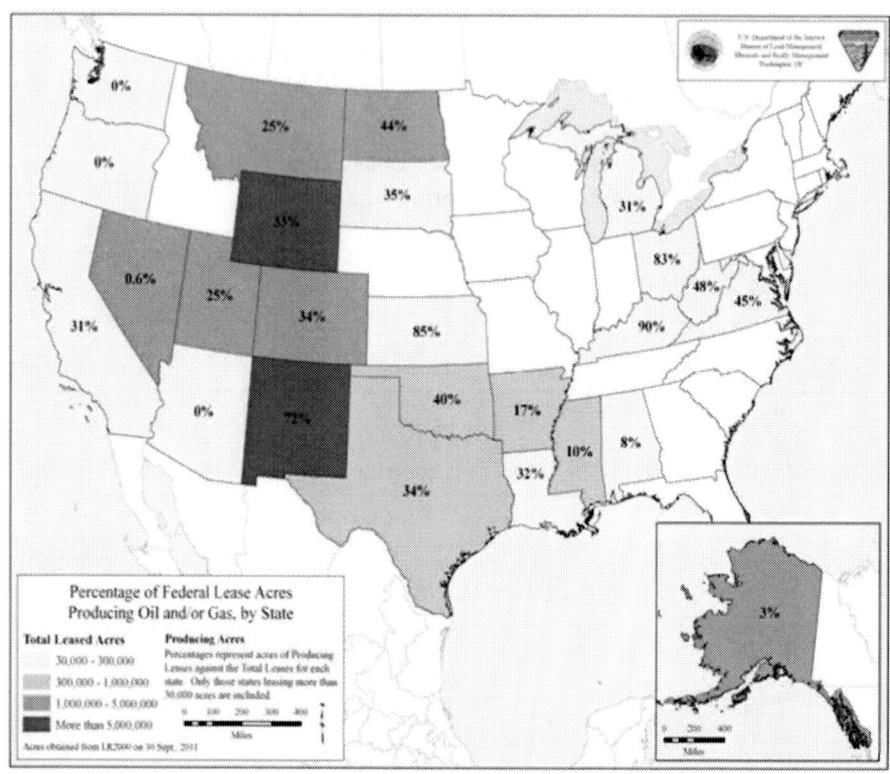

Figure 4. Percentage of Federally-Leased Acres Producing Oil and/or Gas, by State (FY 2011).

Wells in this trend have targeted reservoir rocks of Paleocene to Eocene age and have confirmed the presence of a regionally continuous Lower Tertiary sediment system. The southern part of the Lower Tertiary Trend (in the Keathley Canyon and Walker Ridge areas) is complicated by a salt-canopy system that overlies much of the targeted sediments. Salt canopy thicknesses can vary from 5,000 to 20,000 ft. (1,524 to 6,096 m) in the area, and with water depths ranging from 4,000 to 10,000 ft. (1,219 to 3,048 m), the drill targets can be very deep—25,000 to 35,000 ft. (7,620 to 10,668 m).

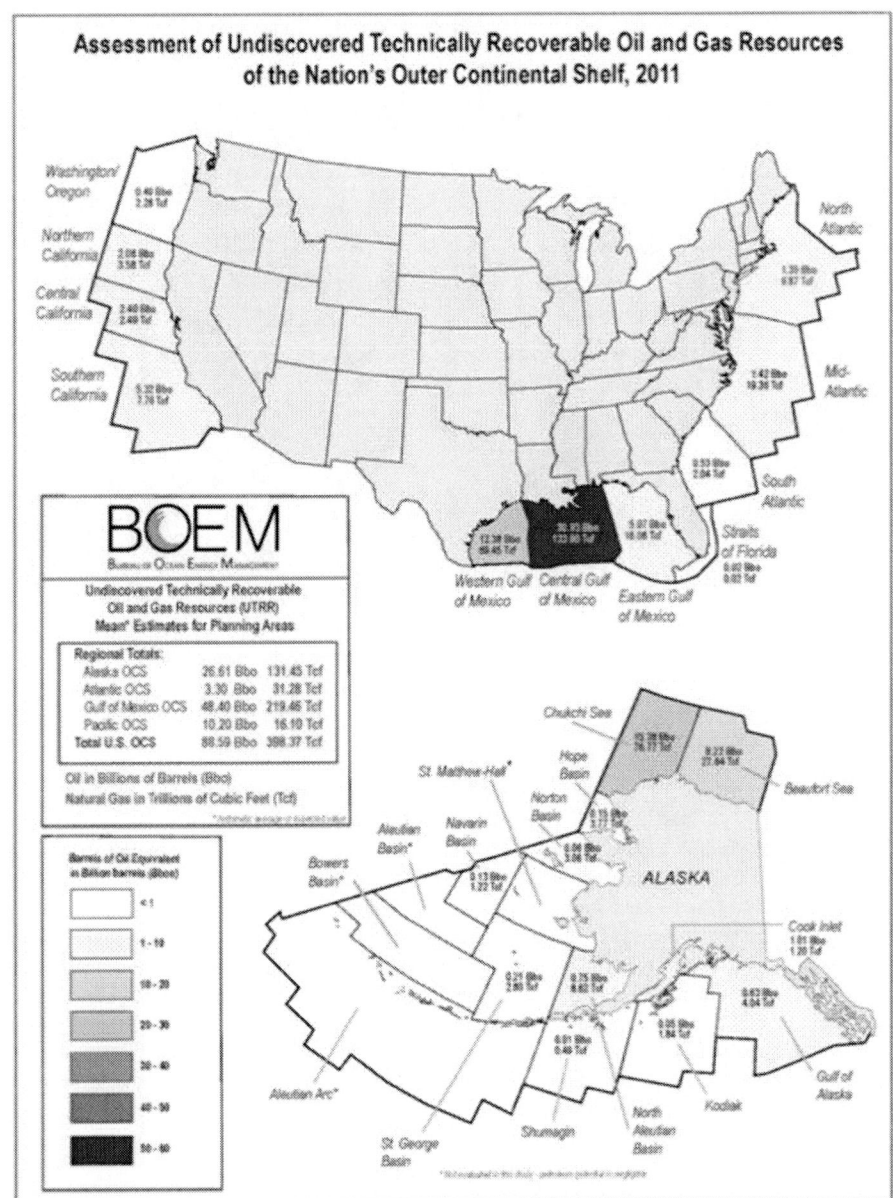

Figure 5. Estimated Undiscovered Technically Recoverable Oil and Gas Resources.

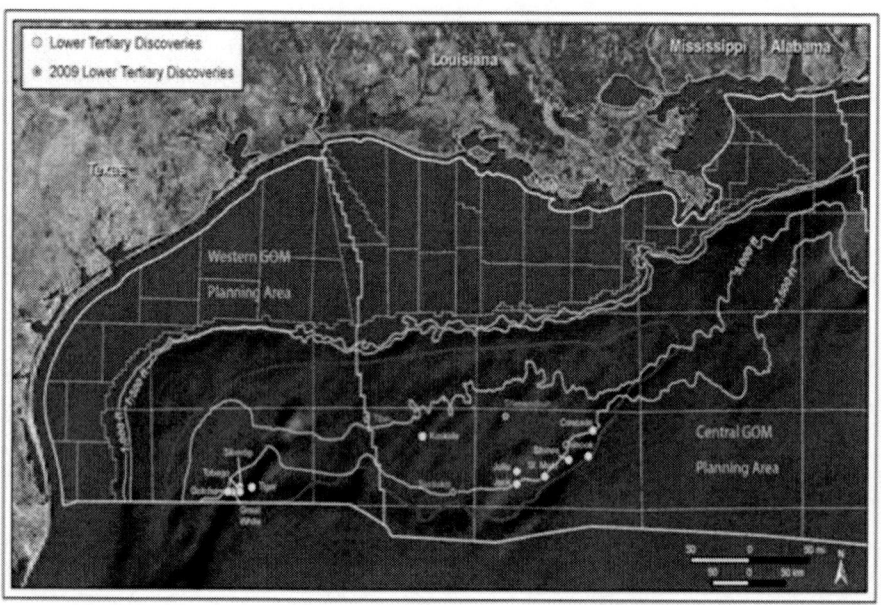

Figure 6. Lower Tertiary Discoveries.

End Notes

[1] Leases in the Central and Western Gulf of Mexico Planning Areas require a "Development Operations Coordination Document (DOCD)."

[2] Onshore Oil and Gas Order Number 1 – commonly known as "Onshore Order #1" – contains the requirements necessary for the approval of all proposed oil and gas exploratory, development, or service wells on all Federal and Indian onshore oil and gas leases. BLM activities with regard to Indian lands do not include issuance of leases and determining their length, rental or royalty, or approval of suspensions or units. These functions are undertaken by the Bureau of Indian Affairs.

[3] Onshore leases vary widely in terms of the number of acres per lease, in contrast to offshore leases which have more uniform size.

[4] Leases in less than 400 meters of water can qualify for a 3-year extension for wells targeting hydrocarbons below 25,000 feet Total Vertical Depth Subsea.

INDEX

A

abuse, 5
access, viii, 8, 11, 14, 19, 29, 30, 42, 49, 65
accountability, 16, 71
accounting, 4, 52
acid, 37
additives, 20, 36, 37
adverse effects, 18, 58
age, 74
agencies, vii, 1, 2, 3, 4, 5, 6, 10, 12, 14, 15, 24, 26, 27, 29, 32, 34, 36, 40, 51, 52
air emissions, 20, 23
air quality, 15, 17, 23, 38
air toxics, 17
Alaska, 48, 53, 60, 61, 62, 64, 68, 71
alienation, 38
ambient air, 15, 17
annual review, 29
ANWR, viii, 42, 45, 49
apples, 51
Arctic National Wildlife Refuge, viii, 42, 45
assessment, 59
assets, 38
atmosphere, 9, 38
audit, 6, 36
authorities, 6, 26, 36
authority, 14, 17, 23, 26, 38, 71

B

bacteria, 37
barriers, 64
base, 63, 72
Beaufort Sea, 60
benefits, 32
BIA, 3, 5, 6, 8, 12, 13, 15, 23, 30, 36, 37, 38, 71
bonuses, 13
Bureau of Indian Affairs(BIA), 3, 5, 76
Bureau of Land Management, ix, 2, 3, 4, 33, 34, 42, 47, 53, 58
burn, 9

C

carbon, 15
carbon monoxide, 15
ceramic, 9
certification, 66
challenges, 5, 28, 29, 40, 47, 52
chemical(s), 16, 19, 20, 21, 36, 37, 39
Clean Air Act, 3, 14
Clean Water Act (CWA), 3, 14
coal, viii, 39, 42, 43
commercial, 21, 48, 52
communication, 9
communities, 38, 39, 69
complexity, ix, 42, 50

compliance, vii, 1, 5, 6, 13, 25, 36, 53
composition, 20, 39
conflict, 70
congress, viii, ix, 38, 41, 42, 43, 45, 48, 49, 52
consent, 38
conservation, 8, 14
construction, 8, 14, 17, 21, 35, 40, 49
containers, 21
contamination, 14, 20
contingency, 21
coordination, 2, 24, 26, 32, 52
corrosion, 37
Court of Appeals, 38
covering, 15, 49, 61, 67
cracks, 37
crude oil, viii, 41, 43, 45, 47, 51
CWA, 3, 14, 18

D

data set, 58
database, 6, 13, 27, 28, 35, 40
Department of Agriculture, 4, 34
Department of Defense, 34
Department of Energy, 39
Department of the Interior, ix, 4, 35, 43, 52, 53, 55
depth, 63, 65, 72
diesel fuel, 17, 39
discharges, 18, 40
disclosure, 19, 20, 39
distribution, 12, 65
District of Columbia, 38
DOI, 43, 48, 53, 57
draft, 2, 18, 34, 39
drainage, 13, 25, 29, 30
drinking water, 14

E

education, 31
EIA, viii, 41, 43, 44, 46, 47, 52, 53
ELGs, 18

endangered, 53
endangered species, 53
energy, vii, viii, ix, 1, 4, 13, 14, 15, 27, 36, 37, 41, 42, 43, 44, 49, 53, 55, 56, 58, 64, 65, 66
Energy Information Administration, viii, 3, 41, 43
Energy Policy Act of 2005, ix, 42, 49
energy prices, 64
energy supply, viii, 41, 43, 44
enforcement, 26, 71
engineering, 71
environment(s), 14, 18, 21, 32, 58
environmental effects, 10, 37
environmental impact, 8, 37, 53
environmental issues, 64
environmental protection, 8, 39
Environmental Protection Agency, 3, 5, 34
environmental quality, 71
environmental standards, 59
EPA, 3, 5, 6, 14, 15, 16, 17, 18, 34, 36, 38, 39
equipment, 9, 10, 17, 21, 23, 38, 52
erosion, 20
evidence, 7, 36, 49, 66
Executive Order, 24, 64
exercise, 18
expertise, 71
extraction, 52

F

federal government, 2, 4, 5, 7, 13, 18, 25, 31, 33, 34, 37, 40
federal law, 14, 34
federal permitting, 51
federal regulations, 15
fiscal year 2009, 6, 27, 28, 31, 35
Fish and Wildlife Service, 4, 6, 34
fluid, 9, 16, 19, 20, 21, 37, 38, 39, 71
food, 31
formation, 2, 6, 8, 9, 10, 11, 20, 25, 35, 36, 37
fractures, 9, 10
fraud, 5

Index

friction, 37
funding, 30, 51

G

GAO, vii, 1, 2, 3, 11, 12, 22, 28, 37, 38, 39, 40
gas development, viii, 2, 4, 5, 6, 7, 10, 12, 14, 15, 16, 17, 18, 19, 20, 21, 22, 23, 24, 26, 27, 29, 30, 31, 34, 35, 36, 37, 38, 42, 43, 45, 47, 51, 56, 58, 60, 66
gas resources, vii, 1, 2, 5, 7, 8, 10, 18, 24, 26, 29, 30, 34, 40, 57, 60
geology, viii, 41, 43
GIS, 4, 29, 30
glycol, 17
governments, 8, 23
grants, 7
greenhouse, 39
greenhouse gas, 39
groundwater, 16, 20, 21, 24
growth, 46
guidance, vii, 1, 2, 5, 13, 14, 15, 16, 17, 19, 24, 25, 26, 27, 28, 29, 32, 33, 34, 35, 37, 39
guidelines, 38
Gulf of Mexico, 44, 46, 48, 53, 56, 60, 61, 62, 63, 64, 65, 72, 73, 76

H

habitat, 14, 53
harvesting, 15
hazardous air pollutants, 17
health, 14, 31, 39
health effects, 39
hiring, 5, 29
history, 64, 67
homes, 22
human, 5, 14, 28, 40, 58, 59, 71
human capital, 5, 28, 40
human health, 14
hydrocarbons, 72, 76
hydrogen, 13

hydrogen sulfide, 13

I

ideal, 10, 11
identification, 18
income, 3
Indian oil, vii, 1, 2, 5, 13, 24, 28, 29, 30, 31, 33, 37, 39
Indian reservation, 7, 38
Indian resources, vii, 1, 5, 9, 12, 13, 25, 26, 29, 30, 32, 33, 34, 35, 37, 38
Indians, 4, 26, 36, 37
industry, ix, 6, 8, 16, 25, 32, 36, 39, 42, 45, 50, 55, 56, 59, 60, 64, 65, 66, 70, 71
infrastructure, 9, 10, 18, 65
inspections, 2, 3, 6, 13, 24, 26, 27, 28, 29, 32, 33, 35, 38, 40, 51
inspectors, 26
integration, 16
integrity, 2, 9, 16, 20, 21
interference, 7
investment(s), viii, 42, 43, 52, 60, 64
issues, 6, 26, 36, 38, 40, 51

J

jurisdiction, 38

L

labor shortage, 52
lakes, 19, 30
laws, 4, 5, 7, 12, 14, 15, 27, 34
laws and regulations, 4, 12, 15, 27
lead, 2, 14, 15, 64, 71
leaks, 17
legislation, 7, 23
light, 23, 59
liquids, 24, 39
litigation, 61, 62, 69
Louisiana, 5, 20, 35

M

majority, 65, 71
management, vii, 1, 2, 5, 6, 13, 14, 16, 18, 21, 22, 23, 24, 27, 29, 34, 35, 36, 37, 38, 51, 52, 53
matter, 15
measurement, 16, 38
memorandums of understanding, 26, 33
Mexico, 5, 27, 35, 57, 60, 63, 64, 65, 73
migration, 20, 68
mineral resources, 37, 58, 71
mission, 14
Montana, 23, 31, 67, 71

N

National Environmental Policy Act (NEPA), 51
National Park Service, 4, 6, 34
national parks, 19, 53
natural gas, vii, viii, 9, 10, 16, 17, 19, 39, 42, 43, 44, 45, 46, 47, 52, 56, 57, 58, 60, 61, 64, 65
natural gas production, vii, viii, 42, 43, 45, 47, 52
natural resources, 70
nitrogen, 15
NPS, 4, 6, 14, 16, 18, 34, 36, 38
nuisance, 22
nursing, 22
nursing home, 22

O

Obama, President, ix, 55
OCS, 56, 58, 60, 65, 72, 73
officials, 3, 5, 6, 13, 16, 17, 18, 19, 20, 23, 25, 26, 27, 28, 29, 30, 31, 32, 35, 36, 39, 40, 64
oil production, viii, 42, 43, 44, 47, 64, 71
oil spill, 59
Oklahoma, 5, 6, 20, 21, 23, 27, 31, 32, 35, 36, 38

operations, 5, 13, 18, 26, 29, 33, 35, 37, 39, 49, 72
opportunities, 7, 33, 39
Outer Continental Shelf (OCS), 52, 56, 60, 64, 65
oversight, vii, 1, 2, 5, 6, 14, 18, 24, 27, 28, 29, 33, 34, 35, 36, 39, 40, 71
ownership, 3, 7, 10, 13, 18, 26, 30
ozone, 15

P

Pacific, 61, 62
parallel, 11, 25
participants, 63
pathways, 36
permeability, 9
permit, 3, 8, 17, 19, 26, 38, 39, 40, 49, 50, 51, 53, 71
Petroleum, 31, 44, 68
pipeline, 23, 24
plants, 17
policy, 35, 44
pollutants, 15, 23, 38
pollution, 16, 23, 39
power generation, viii, 42, 43
preparedness, 60
President, v, ix, 48, 53, 55, 60, 64, 65, 67
President Obama, ix, 55
prevention, 13, 38
primacy, 14
private party, 7
probability, 6, 36
profit, 39
project, 51, 66
property rights, 7
protection, 14, 23, 25, 52, 58, 70, 71
public concern(s), 48
public health, 5, 6, 15, 18, 32, 35
public interest, 39
pumps, 23, 38

Q

quality control, 23
quality standards, 15, 17

R

radius, 21
recommendations, 2, 17, 20, 34, 38
recovery, 4, 5, 10, 23, 25
recreation, 15, 18
recycling, 21
reform(s), 5, 57, 59, 61, 63, 69, 70
regulations, vii, 1, 5, 7, 13, 15, 18, 19, 21, 24, 34, 35, 37, 39
regulatory agencies, vii, 1, 3, 5, 19, 24, 26, 27, 32, 33, 36, 52
regulatory framework, 51
regulatory oversight, 6, 13, 36
regulatory requirements, 17
reliability, 6, 30, 35, 40
Reorganization Act, 7, 37
requirements, viii, 5, 14, 15, 16, 17, 19, 20, 21, 23, 34, 38, 41, 43, 51, 71, 76
reserves, 7, 44, 46
resolution, 65
resource allocation, 40
resources, vii, ix, 1, 2, 4, 5, 7, 8, 9, 10, 12, 13, 14, 18, 19, 23, 24, 25, 26, 27, 29, 30, 32, 33, 34, 35, 37, 38, 39, 40, 50, 52, 56, 57, 59, 60, 63, 65
response, 2, 5, 9, 15, 16, 23, 24, 38, 64
restoration, 14
restrictions, 53
revenue, vii, 1, 2, 4, 5, 25, 30, 31, 36, 71
rights, 5, 7, 10, 12, 18, 30, 51
risk(s), 5, 6, 16, 20, 21, 23, 33, 35, 38, 40
royalty, 3, 12, 13, 31, 32, 63, 70, 72, 76
rules, vii, 1, 2, 5, 6, 10, 15, 16, 18, 19, 20, 21, 22, 23, 24, 25, 32, 33, 34, 35, 36, 39, 40

S

safety, vii, 1, 5, 8, 16, 18, 22, 32, 59, 64, 71
sanctuaries, 52
sandstone rock, vii, 1
school, 22
science, 24
scope, 19, 48
security, viii, 16, 38, 41, 43
sediment(s), 74
seismic data, 65, 72
self-sufficiency, 14
shale, vii, viii, 1, 2, 4, 9, 11, 18, 19, 20, 39, 42, 43, 46
shale gas, viii, 18, 20, 42, 43
shape, 10, 11
shelter, 31
software, 30
South Dakota, 6, 35
staffing, 25
stakeholders, 6, 7, 12, 34, 35, 39
state(s), vii, 1, 2, 3, 4, 5, 7, 8, 10, 12, 14, 15, 16, 17, 19, 20, 21, 22, 23, 24, 26, 27, 29, 31, 32, 33, 34, 35, 36, 37, 39, 40, 51
state laws, 5, 34, 35
statistics, 53
statutes, viii, 37, 41, 43
storage, 9, 17, 21, 38
stormwater, 40
sulfur, 15
sulfur dioxide, 15
suspensions, 63, 76

T

tanks, 9, 21
target, 8, 9, 72
taxpayers, ix, 55, 57, 63
teams, 20, 39
technical comments, 34
techniques, 39, 65
technological advancement, 2
technological advances, 2, 15, 16, 23, 24, 26, 32

technologies, vii, 1, 4, 5
technology, 18, 20, 44, 65
testing, 6, 20, 21, 35, 39
time frame, 3, 31, 33, 34
Title V, 38
tracks, 13
training, 5, 40
transport, 9, 21, 24
transportation, 21
treatment, 10, 18
tribal officials, 30

U

U.S. Army Corps of Engineers, 6, 34
U.S. Department of Agriculture, 6
U.S. Department of the Interior, v, 50, 53, 55
U.S. history, 61
U.S. oil, vii, 43, 47, 52
undiscovered technically recoverable resource (UTRR), 56, 60
uniform, 22, 76
United States, v, viii, 1, 4, 7, 14, 36, 37, 38, 39, 40, 41, 43, 45, 58, 64, 66
updating, 16, 17, 18, 19, 25, 39
USDA, 4, 6, 34

USGS, 58
UTRR, 57, 61

V

vacancies, 28, 40
vessels, 17

W

Washington, 37, 39, 40
waste, 5, 10, 20, 21, 23, 25, 38, 39, 49
waste disposal, 21, 23, 49
waste management, 20
wastewater, 10, 16, 18, 21
water, viii, 9, 10, 13, 14, 16, 17, 19, 20, 21, 22, 23, 24, 36, 37, 39, 42, 48, 57, 63, 72, 73, 74, 76
water quality, 20, 21
welfare, 15, 22
wells, vii, 1, 3, 5, 6, 9, 10, 11, 13, 17, 19, 21, 22, 23, 26, 27, 28, 29, 30, 31, 33, 35, 36, 37, 38, 39, 40, 44, 61, 71, 76
wilderness, 53
wildlife, 14, 18